E, NORTH-EASTERN RAILWAY

GATESHEAD-ON-TYNE, ENGINEER

page 359)

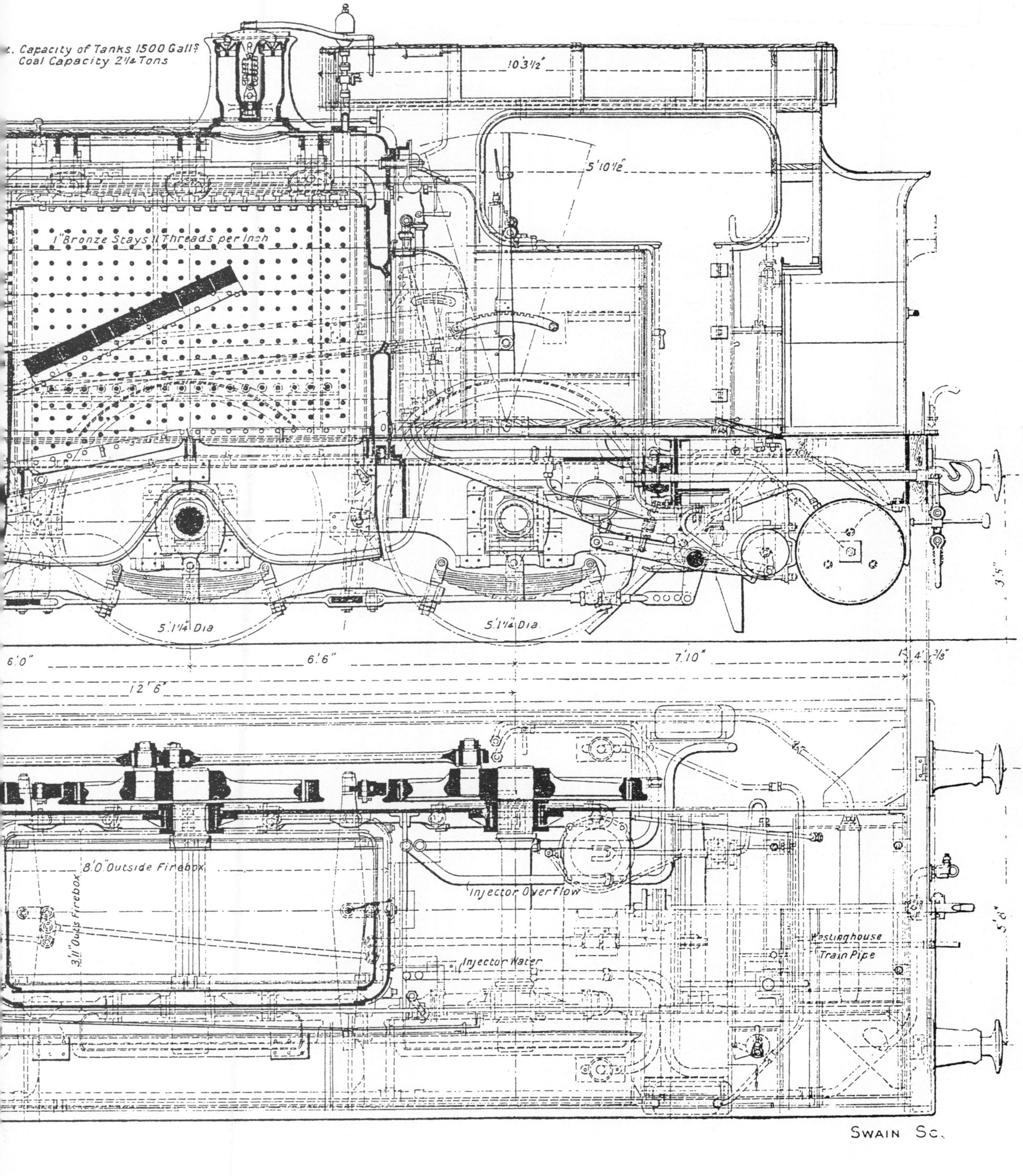

Capacity of Tanks 1500 Galls
Coal Capacity 2¼ Tons

10'3½"

5'10½"

1" Bronze Stays 11 Threads per Inch

3'5"

5'1¼" Dia.   5'1¼" Dia.

6'0"   6'6"   7'10"   4' 3/8"

12'6"

8'0" Outside Firebox

3'11" Outs Firebox

Injector Overflow

Injector Water

Westinghouse Train Pipe

5'8"

Swain Sc.

Volume 3

North Eastern Record

A survey of the Locomotives of the
North Eastern Railway

Prepared for
The North Eastern Railway Association
by
J M Fleming

Published by
The Historical Model Railway Society

This Volume is dedicated to

J S MacLean

and my Friends in the

North Eastern Railway Association

Companion with this Volume:

North Eastern Record - Volume 1   ISBN 0 902835 13 0

North Eastern Record - Volume 2   ISBN 0 902835 19 X

First Published 2000.

ISBN 0 902 835 20 3

All rights reserved. No part of this book may be reproduced or transmitted in any form or by any means, electronic, mechanical, including photocopying, recording or by any information storage and retrieval system, without written permission from the Publisher.

© 2000 John Mansell Fleming and The Historical Model Railway Society.

Book design by Trevor Johnson.

Printed in Great Britain by The Amadeus Press Ltd..

Volume Three Contents

	Introduction	iii
	Dimensions, Drawings and Liveries	iv
Chapter 1	The Origins and Early Years of the North Eastern Railway	1
Chapter 2	Amalgamations 1862 - 1865	30
Chapter 3	NER Locomotive Progress under Two Locomotive Superintendents, 1863 - 1869	44
Chapter 4	NER Locomotive Progress under Two Locomotive Superintendents, 1870 - 1876	54
Chapter 5	Locomotive Development 1877 - 1882, Edward Fletcher, Locomotive Superintendent	72
Chapter 6	Locomotive Development 1883 - 1885, Alexander McDonnell, Locomotive Superintendent	85
Chapter 7	Locomotive Progress 1886 - 1890, Thomas William Worsdell, Locomotive Superintendent	91
Chapter 8	Locomotive Progress 1890 - 1910, Wilson Worsdell, Locomotive Superintendent	108
Chapter 9	Locomotive Progress 1910 - 1922, Vincent Litchfield Raven, Chief Mechanical Engineer	132
Chapter 10	Locomotives for the Officers' Special Saloons	144
Appendix 1	North Eastern Railway Locomotives in Preservation	154
Appendix 2	The Duplicate List	155
Appendix 3	Wartime Loans	156
Appendix 4	Some Dimensions	158
	Index	159

In 1894 David Jones pioneered the use of 4-6-0s in Britain with his Highland Railway goods locomotives. Wilson Worsdell introduced the first British express passenger 4-6-0 with the North Eastern Railway's Class S of 1899, soon followed by the larger-wheeled Class S1. No 2111 was the first locomotive of Class S1, completed at Gateshead Works in December 1900 and seen here in the grey paintwork adopted by the NER for official photographs. The lining corresponds to that of the original green working livery. The Company's large heraldic device is seen on the tender, with the small version on the splasher of the middle coupled wheel, just above the barely discernable hump provided to clear the big end of the connecting rod.

Introduction

This third volume in the series *North Eastern Record*, portraying the image of the North Eastern Railway, looks at locomotives and their liveries. The story begins with three companies which formally amalgamated to form the NER in 1854, while others which joined later are also covered. There are two exceptions to this. The Stockton & Darlington Railway has recently been the subject of a detailed locomotive history, *The Locomotives of the Stockton & Darlington Railway* by Tom Pearce, published by HMRS in 1996, while the Hull & Barnsley Railway, which united with the NER in 1922, just before 'Grouping', has been covered by Ron Prattley's *Locomotives of the Hull & Barnsley Railway*, published by HMRS in 1997, and Martin Barker's *An Illustrated History of Hull & Barnsley Locomotives, Volume 1*, published by Challenger Publications in 1996. The Hull and Barnsley has therefore been excluded from this volume, while the story of S&D locomotives has been to some extent restrained.

The early period of NER locomotive history presents a number of problems, exacerbated by three features which were not peculiar to the company:

1. Although the *Whyte* wheel arrangement classification has been adopted throughout this book, this was not introduced until 1900 and early descriptions often refer to the number of coupled and carrying wheels without specifying the layout. I have restricted the term 'driving wheels' to those driven directly by a connecting rod; coupled wheels include these and any others driven indirectly by coupling rods.

2. The classification of NER locomotives was introduced by T W Worsdell in 1886 and applied retrospectively to those of his predecessors and surviving relics of the constituent companies. The classes thus imposed on locomotives built up to the end of Edward Fletcher's regime, in 1882, were in many cases an attempt to impose order on apparent chaos and often group together locomotives of widely diverse characteristics.

3. Once an initial stock had been built up, allocation of numbers to new locomotives did not follow the numerical sequence with which we are now familar. Instead, many new locomotives acquired numbers from old ones which were being withdrawn or, at times, sidetracked on to a duplicate stock list. This reflected accounting conventions whereby new construction could be charged either to 'Capital', if it reflected an expansion of stock to cope with growth in routes or traffic, or 'Revenue', if it was a replacement, albeit a more powerful one. In practice, the NER charged much new locomotive construction to Revenue and allocated old numbers to new locomotives, thereby splitting classes into a mixture of numbers in sequence and others dotted around.

My own interest was stimulated by earlier researchers, notably J S MacLean, R H Inness, and E L Ahrons. J S MacLean published the first book devoted to NER locomotives in 1905, and followed it up in 1925 with a revised and more detailed work. Though superseded by later research, this remains an attractive account and it has been a pleasure to include many of his original drawings in this volume. I have also had access to his papers and correspondence. MacLean was acquainted with many people at Gateshead Works, when it was the centre of NER locomotive design and construction. One of these was R H Inness, who began at Gateshead in October 1898, entering the locomotive drawing office in 1905 and transferring with the other design staff to Darlington in 1910; latterly in charge of the locomotive drawing office there, he was Head Locomotive, Carriage & Wagon Draughtsman at his retirement in 1949. He did extensive research into NER locomotive history, and a number of the resulting drawings have been used here.

I am also glad to have been able to include some drawings by the late Laurie Ward, who practised his skills in the London County Council Architect's Department and was one of the early stalwarts of the NERA. Most of the drawings are, however, my own. The photographs also, unless otherwise credited, are from my collection.

The late Ken Hoole was a prodigious researcher into most aspects of NER history, and his last book, published in 1988, was *An Illustrated History of NER Locomotives*. This is in many ways complementary to the present work, particularly since he restricted his attention to locomotives built from 1883 and such earlier classes as survived until 1894.

Getting a book to press can be arduous, and I should very much like to thank the 'production team' of North Eastern Railway Association colleagues: Bill Fawcett, Geoff Horsman, John Mallon (who has also contributed extensively to Chapter 1), John Richardson, John Addyman and John Proud for their efforts, together with Trevor Johnson of the Historical Model Railway Society. Others to whom I owe thanks include John Edgington, Colin Foster, the late Arthur Nunn, the late Ken Taylor, Gerry Young, and Ann Wilson of the Darlington Railway Centre and Museum.

J M Fleming
January 2000

Dimensions, Drawings and Liveries

Throughout this book, dimensions are quoted in imperial units: those in which the locomotives were conceived and built. Thus lengths are given in feet (ft) and inches (in); capacities in gallons and tons; pressures in pounds (force) per square inch (psi). Linear dimensions are normally quoted to the nearest ¼in, but the reader should bear in mind that the tolerance on some measurements can be greater than this; for example, some sources appear to quote wheel diameters across new tyres, others across tyres with average wear, without specifying which convention they have adopted.

For modelling locomotives it would be ideal if all scale drawings were presented in the form of front and side elevations, cab views and details of the differences between the two sides. In some cases front elevations have been provided but, in accordance with practice at the time they were first drawn, the majority of drawings are confined to a single side elevation. The photographs include many three-quarter views, from which further details can be deduced, and Appendix 4 lists key transverse dimensions of most of the later locomotives. Throughout Chapters 2 to 10, scale bars have been provided on nearly all drawings but this practice has not been adopted in Chapter 1, where one cannot always be so convinced of the reliability of the original source; these drawings can, nonetheless, be scaled from the wheel dimensions given. The subject of tenders requires special attention, although there was a considerable degree of standardisation, and it is intended that this be covered by a separate, short work.

Colours are notoriously difficult to recall accurately, and are rarely mentioned in official records. Thus the only contemporary sources are coloured drawings and paintings, such as Colour Plate 1. Even if these were accurate and have survived without significant colour change, both bold suppositions, one cannot be sure how they will appear when printed. Likewise, contemporary colour plates of the later locomotives, such as those published by the *Railway Magazine*, are no more than a very approximate guide; for instance the *Railway Magazine's* rendition of North Eastern green was known by contemporaries to be far too dark a shade. Thus the colour plates presented here are not to be regarded as an infallible guide. The most reliable sample, Colour Plate 3, is a plate painted in 1925 by the painter who was working on the restoration of the first *Tennant*, No 1463 (now exhibited at Darlington), and which has been stored away from strong light since then.

For almost a century an error has been perpetuated with regard to the colour of outside frames on locomotives during Edward Fletcher's term as Locomotive Superintendent. At different periods these were either green or a very dark red, but an article in *Locomotive Magazine* for 15 April 1910 erroneously gave the colour as a 'light Indian red'. This arose from a mistaken transcription of a letter originally sent to John Brown, who later persuaded the LNER to preserve *Tennant* No 1463. This letter referred to the *insides* of the frames as being Indian red.

A Select List for Further Reading

This book has been based on extensive research into original sources, without reliance on other published material. For those seeking further information about specific locomotives or engineers, however, the following publications may prove useful.

Ken Hoole	*An Illustrated History of NER Locomotives*	OPC, 1988
Railway Correspondence & Travel Society	*Locomotives of the LNER*	
Geoffrey Hill	*The Worsdells – A Quaker Engineering Dynasty*	Transport Publishing Co, 1991

North Eastern Express, the journal of the North Eastern Railway Association, publishes articles on individual locomotives and classes.

Chapter 1

The Origins and Early Years of The North Eastern Railway

The Constituent Companies.

The North Eastern Railway (NER) was formed in 1854 by the amalgamation of three existing companies: the Leeds Northern Railway (LNR), York & North Midland Railway (Y&NMR) and York, Newcastle & Berwick Railway (YN&BR), of which the latter was the largest.[1] However, the YN&BR was itself the product of several earlier amalgamations and it is convenient to start this record with its direct ancestor, the Newcastle & Darlington Junction Railway, although this was not the earliest of the companies included in the YN&BR.

The Newcastle & Darlington Junction Railway (N&DJR)

The N&DJR was opened on 18 June 1844 but had no rolling stock of its own until the end of that year, the services being worked initially by locomotives and rolling stock provided by the adjoining Great North of England Railway (GNER), which had opened to mineral traffic on 4 January 1841 with passenger services starting on 30 March. On 27 July 1846 the N&DJR purchased the GNER to form a combined company, the York & Newcastle Railway (Y&NR).

[1] The amalgamation Act provided for the inclusion also of the modest Malton & Driffield Junction Railway, which had no locomotives of its own and joined the NER from 28 October 1854.

The first locomotives acquired by the N&DJR were four 2-2-0s, built in 1839-40 for the Midland Counties Railway and purchased from the Midland Railway in 1844; these had 4ft 0in diameter leading wheels and 5ft 6in driving wheels. Nine long-boilered 2-4-0s with 'haystack' fireboxes (Fig 1.1) were built in 1844-5 and had 3ft 8½in leading and 5ft 8in coupled wheels with 14in by 22in outside cylinders.

For goods and mineral traffic seven 0-6-0s were delivered during 1844-6 with 4ft 8in diameter wheels and 15in by 24in cylinders. Six more, having the same dimensions and built in 1845 (Fig 1.2), originally had a wide footboard, the full length of the frames, with a handrail on its outside. Two mixed-traffic 0-6-0s were also supplied, with 5ft 0in wheels.

Anticipating an earlier opening of its line than proved possible, the GNER placed extensive orders in 1839, on the advice of James I'Anson Cudworth, its Locomotive Superintendent until his move to the South Eastern Railway in 1845. Ten 2-2-2s, for use on passenger trains, were supplied in 1839, with 3ft 6in carrying wheels and 5ft 6in driving wheels; two more followed in 1840. The majority were built

Fig 1.1
York, Newcastle & Berwick Railway No 26, a long-boiler locomotive built by Robert Stephenson & Co for the N&DJR in 1845. *(J S MacLean)*

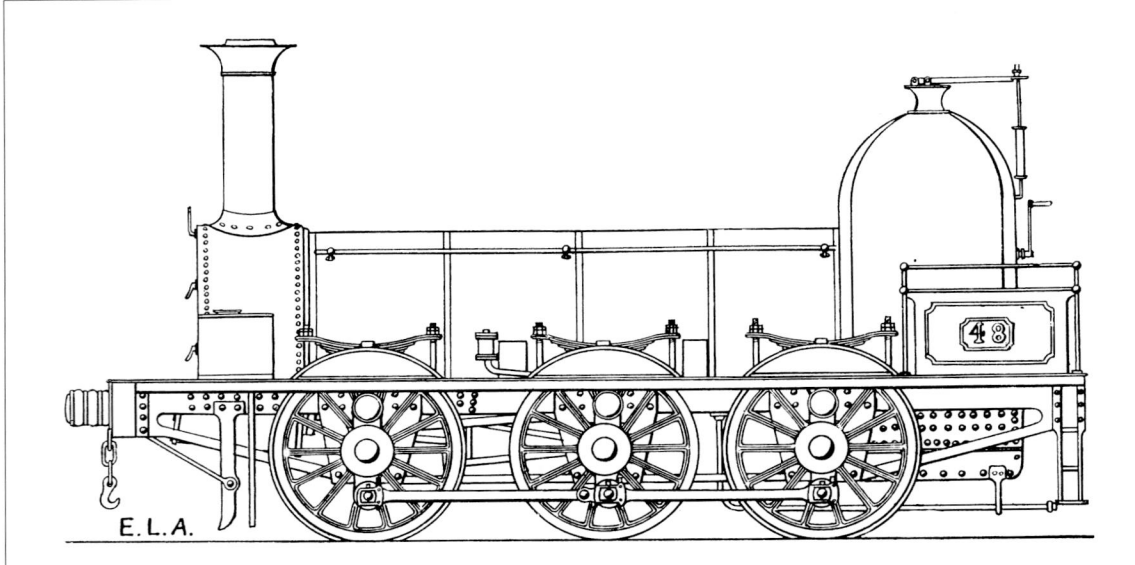

Fig 1.2
York, Newcastle & Berwick Railway No 48, an 0-6-0 goods locomotive built by R B Longridge & Co for the N&DJR in May 1845. *(E L Ahrons)*

Fig 1.3
York, Newcastle & Berwick Railway No 66, *Richmond*, built by R & W Hawthorn, is shown as rebuilt by them in 1849. It was given new cylinders 15½in by 20in. *(J S MacLean)*

by R & W Hawthorn, who supplied a further express 'single', *Richmond*, in 1845. This is said to have achieved an average speed of 56mph in a test run with two carriages on the generally straight and gently-graded GNER main line. It had 3ft 6in and 6ft 6in wheels and 16in by 21in cylinders, with special expansion gear in addition to link motion. Following an accident, it was rebuilt in 1849 (Fig 1.3).

The GNER route was used extensively for trials while the Gauge Commissioners were comparing the merits of the Stephenson gauge and Brunel's broad gauge. To demonstrate the potential of the 'narrow' gauge, in 1845 Robert Stephenson & Co delivered to the GNER for trials a 4-2-0 intended for the N&DJR and identified only by the letter 'A' (Fig 1.4). It had 3ft 6in carrying wheels, 6ft 7in diameter driving wheels and a total wheelbase of only 12 feet. An oval boiler was adopted, to keep down the centre of gravity, and was 13ft 6in long with a working pressure of 90psi, supplying cylinders 15in by 24in. Having proved its worth, the 'Great A' remained on the GNER.

For goods and mineral traffic the GNER initially purchased four-coupled locomotives. Seventeen were supplied in 1839, the majority from Hawthorn's and mostly with 3ft 6in carrying wheels and 4ft 6in coupled wheels; they are thought to have been 2-4-0s.[2] A final 2-4-0, built in 1845, had 5ft 6in coupled wheels and is illustrated in Fig 1.5. One 0-6-0 had been purchased in 1839 (having 4ft 6in wheels), but the move towards this type only began in 1844 when Stephenson's supplied two larger ones with 4ft 8in wheels; a final four 0-6-0s, with 4ft 6in wheels, were delivered by Hawthorn's in 1846.

Although no record of the earliest liveries has been found it is almost certain that the locomotives of the N&DJR and GNER would be painted green, which was the colour subsequently adopted by the Y&NR. The N&DJR locomotives were numbered, but those of the

---

[2] Although R H Inness suggested they had been 0-4-2s, with some possibly being rebuilt to 2-4-0s

Fig 1.4
Robert Stephenson & Co's 4-2-0 express locomotive 'A' of November 1845, seen as York, Newcastle & Berwick Railway No 38. *(J S MacLean)*

GNER only carried names until they were incorporated into the Y&NR stock.

Meanwhile, on 1 September 1844, the N&DJR had acquired the Brandling Junction Railway (BJR), the purchase being authorised by Parliament in 1845. The first part of the BJR opened on 15 January 1839 but four locomotives had already been acquired since March 1838. Of these, two were 2-2-2s, supplied for passenger traffic by R B Longridge & Co; both had 3ft 6in carrying wheels whilst one had 5ft 0in driving wheels and the other 5ft 6in. The other 1838 purchases were 0-6-0s; one from Longridge's with 4ft 6in wheels and the other from Hawthorn's with 4ft 0in wheels. Two further 0-6-0s with 4ft wheels were supplied by the same firms in 1839, at the same time as four 2-4-0s built by R & W Hawthorn. Two of the 2-4-0s had 4ft 0in carrying wheels and 5ft 0in coupled wheels, the others had 3ft 6in and 4ft 6in wheels. All ten locomotives were named and were provided with four-wheeled tenders carrying 800 gallons of water and one ton of coke.

The livery of the BJR locomotives was maroon or brown, except for the 0-6-0 named *Brandling* which was painted red.[3] The BJR Revenue Account for 1843 includes a reference to the purchase of Brunswick green, vermilion and chrome yellow paints in the lists of materials "used for the repair of engines"; a mixture of green and vermilion would produce maroon or brown, the relative proportions of the two components determining the resulting colour.

Fig 1.5
York, Newcastle & Berwick Railway No 75, a 2-4-0 built by Robert Stephenson & Co for the GNER in July 1845. It had cylinders 15in by 22in. *(J S MacLean)*

[3] Tomlinson, W W, *The North Eastern Railway - Its Rise and Development*, p 394

The York & Newcastle Railway (Y&NR)

For passenger traffic the Y&NR acquired only 2-4-0 locomotives, a total of 37 being purchased in no fewer than nine lots. Robert Stephenson & Co built ten in 1847 (Fig 1.6) with 3ft 6in carrying and 6ft 0in coupled wheels and 15in by 22in cylinders; six-wheeled tenders held 1,200 gallons of water. Three more, of an improved design for steadier running at high speed, were also built in 1847 (Fig 1.7 and Colour Plate 1), with wheel diameters increased to 3ft 8in and 6ft 1in, and outside cylinders of the same dimensions as the first batch. These were nicknamed the 'cross-legged' locomotives because of the positions of the connecting and coupling rods when in motion. Their boiler pressure was 100psi, and two pumps worked off the driving axle to feed water to the boiler. Each tender had six brake blocks, working on the left-hand wheels only. The other twenty-four 2-4-0s had coupled wheels ranging in diameter from 4ft 6in to 6ft 0in diameter and equally varied cylinder dimensions.

The Y&NR employed 0-6-0s on goods and mineral traffic, all having cylinders 15in by 24in. The first six had 4ft 9in wheels, while most of their successors had 4ft 8in wheels – 31 locomotives of this size being supplied to the Y&NR and YN&BR down to June 1851.

On 12 October 1846 the Y&NR took possession of the Hartlepool Dock & Railway (HD&R) and the Great North of England, Clarence & Hartlepool Junction Railway (GNEC&HJR), Parliamentary approval for amalgamation being received on 22 July 1848. The coal traffic on these lines was worked originally by the colliery companies' own locomotives, some 20 of which were subsequently taken into Y&NR stock. In 1840 the HD&R had bought four small second-hand locomotives and one new one for the 'coach trains', these being followed in 1841 by another new 'coaching engine' of the "same power and size as the preceding one but to be black instead of polished". Evidently the HD&R did not consider that colour or ornament was important as the brasswork from all six locomotives was sold two years later.

Of the 0-6-0s used by the HD&R on coal traffic, twelve had 4ft 0in wheels in common but seven different cylinder sizes, while five had

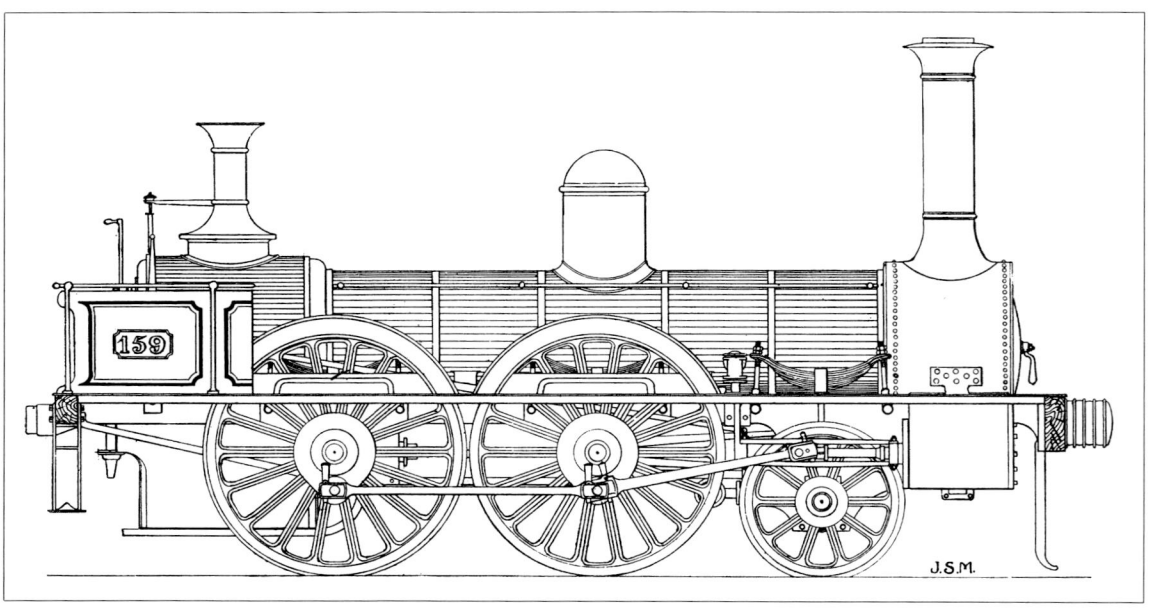

Fig 1.6
York, Newcastle & Berwick Railway No 159, a Stephenson long-boiler 2-4-0, built in October 1847. (*J S MacLean*)

Fig 1.7
York, Newcastle & Berwick Railway No 144, a 'cross-legged' express locomotive, built by Stephenson's in July 1847. (*J S MacLean*)

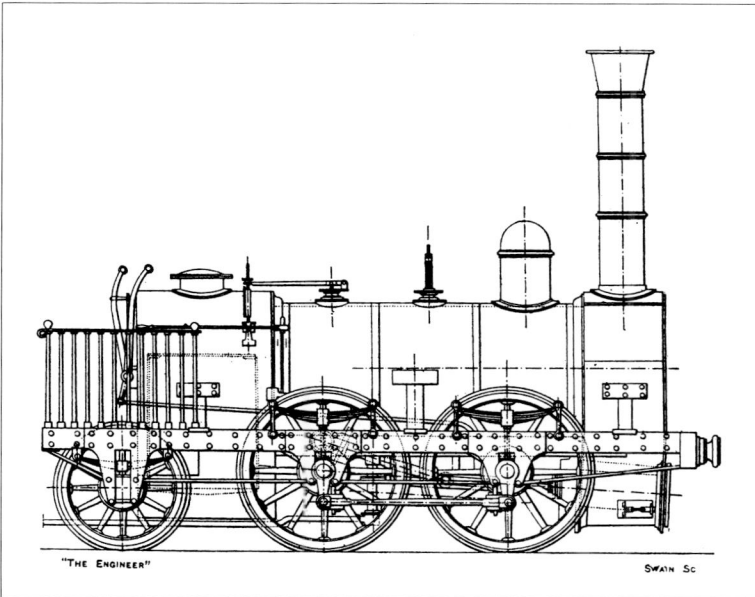

Fig 1.8
Stanhope & Tyne Railway, Stephenson's 0-4-2 locomotive, supplied in 1834. It had 4ft 6in and 3ft 6in wheels, and cylinders 14in by 18in. *(John Swain)*

4ft 6in wheels. Goods and mineral traffic 0-4-0s had 4ft 0in wheels. Originally, the locomotives were named but probably not numbered.

The Y&NR took possession of two more lines on 1 January 1847, these being the Pontop & South Shields Railway (P&SSR) and the Durham & Sunderland Railway (D&SR). The D&SR was worked by stationary engines and ropes until after the formation of the NER.

The Pontop & South Shields Railway originated as the eastern section of the Stanhope & Tyne Railway (S&TR) which had opened in 1834, when it acquired an early 2-2-0 of Stephenson's *Planet* type, having 3ft 6in and 5ft 0in wheels and 11in by 16in cylinders. Unusually for this period, the S&TR also built one locomotive at its own workshops in South Shields; the details of this remain unknown. The company's business was coal and mineral traffic, for which they relied chiefly on 0-4-2s, turning to Stephenson's for the majority of the five supplied in 1834 (Fig 1.8). A similar one was obtained from Hackworth's in 1837 and another from Stephenson's in 1838, though it had a slightly longer boiler and much larger heating area. The S&TR's first 0-6-0 was delivered in 1837, and they finished up with three: two with 4ft 0in wheels and one with 4ft 6in wheels. The locomotives of the P&SSR were named and numbered, their livery being slate colour.[4]

The York, Newcastle & Berwick Railway (YN&BR)

The northern section of the YN&BR originated as the Newcastle & Berwick Railway (N&BR), the first portion of which was opened on 1 March 1847 between Heaton and Morpeth; the final mile and a half from Heaton to a temporary terminus in Newcastle was over

---

4 Tomlinson, *op cit* pp 394-5

part of the existing Newcastle & North Shields Railway (N&NSR) which was forthwith incorporated in the N&BR. The N&BR was opened throughout from Newcastle to Tweedmouth on 1 July 1847 and was amalgamated with the Y&NR on 8 August 1847 to form the YN&BR; the two sections of the new company were not actually joined until the completion of a temporary bridge across the River Tyne in August 1848.

The Newcastle & North Shields Railway opened on 22 June 1839 but had purchased its first locomotive second-hand in 1838 from the Leeds & Selby Railway; this was one of their original 2-2-0s of 1834. Of the later N&NSR 2-2-2s, one had 6ft driving and 4ft 0in and 4ft 6in carrying wheels whilst two others had 5ft 6in driving and 3ft 6in carrying wheels. The railway was, unusually for this area, essentially a passenger and goods carrier, excluded from any profitable share in the coal-shipping traffic by a clause in its Act.

A solitary N&NSR 2-4-0 had 4ft 9in coupled wheels with 3ft 6in carrying wheels and 14in by 18in cylinders, while an 0-4-2 built in 1839 had wheels and cylinders of the same dimensions but a longer and broader boiler. Another 0-4-2 having wheels and cylinders of the same size was built by R & W Hawthorn in 1840 and is notable for having one of the earliest known examples of a 'steam dryer', the precursor of the superheater. It had a return-tube boiler, with the chimney at the footplate end (Fig 1.9), while the firebox was in two parts with a water space between them. Steam passed to the cylinders through a 'steam chamber' in the upper part of the smokebox, while the flue gases reached the chimney through a number of vertical tubes passing through the steam chamber.

The York Newcastle & Berwick Railway reverted to the 2-2-2 type for its passenger locomotives. Two, built in 1847, had 6ft 6in driving wheels, while *Plews*, built in 1848 (Fig 1.10) and named after

Fig 1.9
Newcastle & North Shields Railway. R & W Hawthorn's superheated 0-4-2, built in February 1840. *(John Swain)*

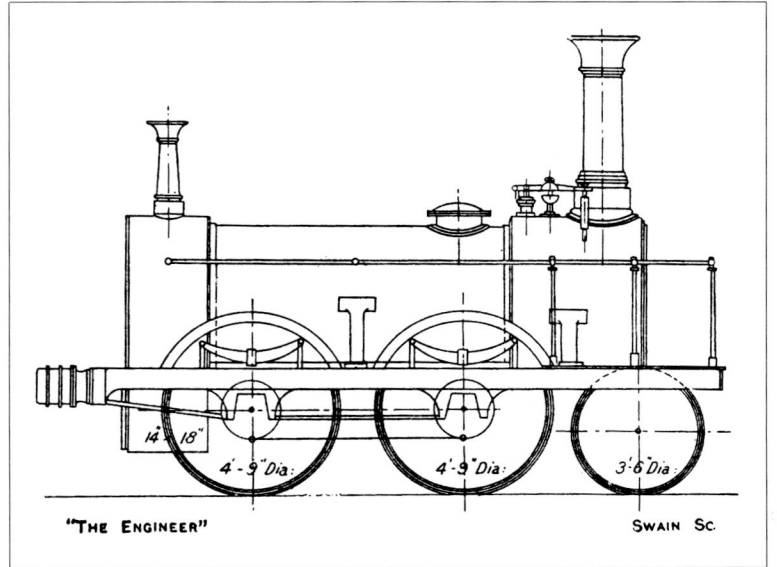

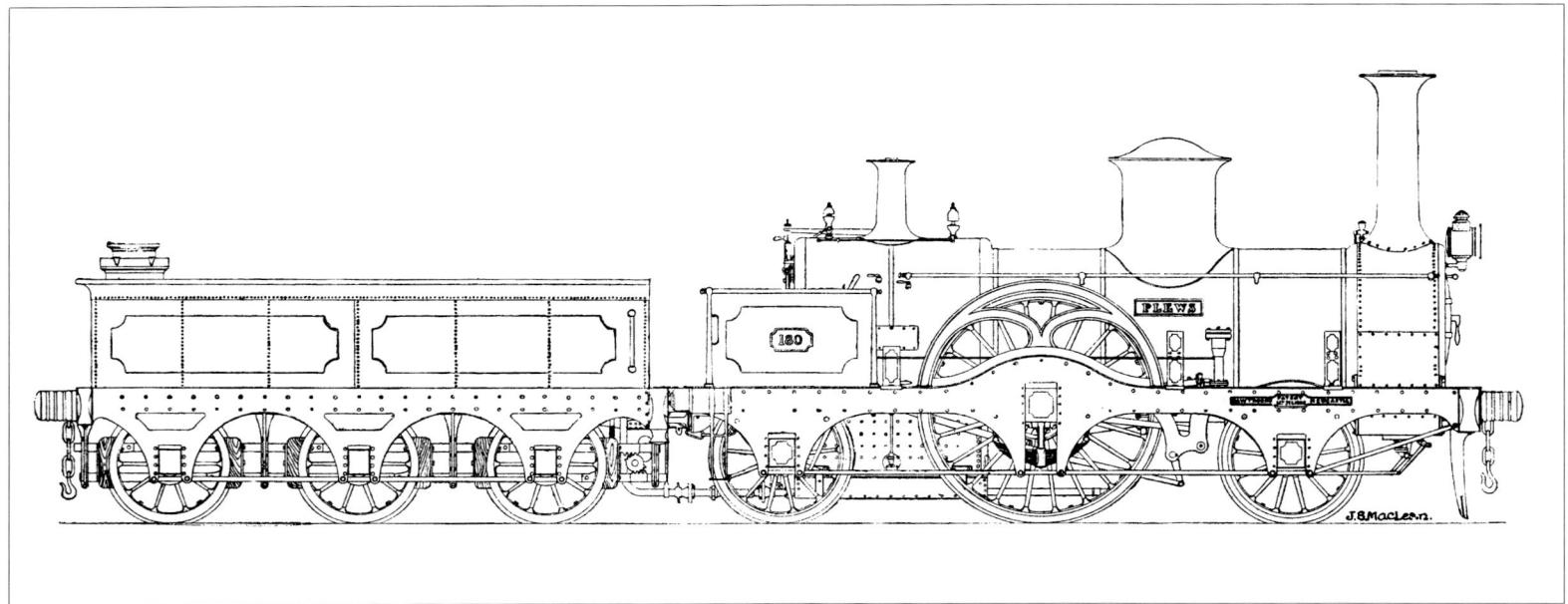

Fig 1.10
York, Newcastle & Berwick Railway No 180, *Plews*, built in December 1848 by R & W Hawthorn & Co. *(J S MacLean)*

Nathaniel Plews, a director, had 4ft 0in carrying wheels and 7ft 0in driving wheels. An oval boiler working at 120psi supplied steam to 16in by 20in cylinders. Its tender had 3ft 6in wheels and a water capacity of 1,400 gallons. Hawthorn's supplied an apparently identical locomotive, *The Queen*, to the North British Railway to haul East Coast expresses north of Berwick. In 1849 Robert Stephenson & Co delivered another celebrated express locomotive, No 190. This had 3ft 10in carrying and 6ft 7in driving wheels, an oval boiler and 16in by 20in cylinders. The eccentrics were outside the driving wheels (Fig 1.11) and the sheaves lay beyond on extensions of the driving axle. Its metamorphosis, through several rebuildings, is told in Chapter 10.

One 2-2-2 of 1853 was similar to No 190, though with 4ft 0in carrying and 6ft 3in driving wheels, while the same year brought an unusual rebuilding in which a 2-4-0 of 1847 became another 2-2-2 (Fig 1.12); its 6ft 6in driving wheels had slightly curved spokes but the cylinders remained as before. An acquisition of 1854 was *Sanspareil*, which had been built by Hackworth's in 1849; it had 4ft 0in carrying wheels and 6ft 6in wrought-iron driving wheels, with only ten spokes. Originally, the slide valves allowed part of the steam needed for the return stroke to enter the cylinder before the forward stroke was completed, so as to form a cushion between the piston and cylinder covers. It was one of the earliest locomotives to have its boiler covered with sheet-iron cleading (this being the term used in the region to denote boiler cladding) (Fig 1.13).

In 1846 Stephenson and Howe patented their system of balancing locomotives, which eliminated the alternate vertical thrust by the connecting rods of two-cylinder locomotives. Two 4-2-0 locomotives were built with this three-cylinder 'patent' arrangement (Fig 1.14), having 3ft 6in carrying and 6ft 6in driving wheels. The inside cylinder was $16^{3}/_{8}$in by 18in and the two outside cylinders were 10½in by

Fig 1.11
York, Newcastle & Berwick Railway No 190, Stephenson's famous 'single' express locomotive of January 1849. In a test run it hauled 13 carriages while consuming only 18lbs of coke per mile. *(J S MacLean)*

Fig 1.12
York, Newcastle & Berwick Railway No 159 was rebuilt from a 2-4-0 (Fig 1.6) in December 1853. *(J S MacLean)*

Fig 1.13
York, Newcastle & Berwick Railway No 135, *Sanspareil*, was built by Timothy Hackworth at Shildon in December 1849 but not purchased until 1854. *(J S MacLean)*

Fig 1.14
York, Newcastle & Berwick Railway No 77 was one of the three-cylinder 4-2-0-s built to Stephenson & Howe's patent, this one in November 1846 and No 102 in February 1847. *(J S MacLean)*

Fig 1.15
York, Newcastle & Berwick Railway No 77 as rebuilt by Robert Stephenson & Co in 1852-3. (J S MacLean)

22in. These became YN&BR Nos 77 and 102, and in 1852-3 No 77 was rebuilt as a 2-2-2 (Fig 1.15), retaining its three cylinders; the new wheel diameters were 3ft 9in and 6ft 8in. The slide valves were operated by two sets of link motion, one for the inside cylinder and one for the outside pair through a transverse shaft. Two more 4-2-0s, developed from the 'A' of 1845, were built by Stephenson's in 1847 (Fig 1.16) with wheels 3ft 6in and 6ft 6in diameter and cylinders 15in by 24in. Intended for an Italian railway, they were purchased instead by the YN&BR, becoming Nos 147 and 148, and the latter was rebuilt as a 2-2-2 with outside bearings for the carrying wheels; its frames were wide and partly enclosed the original outside cylinders which had been retained.

The building of 2-4-0s for the YN&BR began in September 1847 with two having 4ft 8in coupled wheels. A further two supplied that year had 3ft 6in carrying and 5ft 0in coupled wheels, the leading pair of these being flangeless; the cylinders were 14in by 18in and the tenders had 3ft wheels. One of these was built by J Coulthard of Gateshead (Fig 1.17) and carried the name *Jenny Lind* - more usually associated with the 2-2-2s built by E B Wilson & Co of Leeds. Seven 2-4-0s built by Hawthorn's in 1847-9 had 3ft 6in carrying and tender wheels, 5ft 6in coupled wheels and 15in by 22in cylinders (Plates 1.1 and 1.2; Fig 1.18). Two pumps near the firebox, worked from eccentrics placed between the cranks on the driving axle, fed water to the boiler which had a working pressure of 80psi. The trailing axleboxes had 'grasshopper' springs below them.

Although not delivered until 1848-9 five 2-4-0s had been ordered from Robert Stephenson & Co in 1846 with 3ft 7in and 6ft 1in wheels and 15½in by 22in cylinders; equipped with 'haystack' fireboxes they looked antiquated. Outside valve gear was fitted, the eccentrics being placed between the outside frames and the leading coupled wheels which were flangeless (Fig 1.19). Another 2-4-0 built in October 1848 had 5ft 6in coupled wheels whilst a much smaller one of 1849 had

Fig 1.16
York, Newcastle & Berwick Railway No 147 in its original form as a Stephenson & Co 4-2-0, acquired by the YN&BR in August 1847. (J S MacLean)

Fig 1.17
York, Newcastle & Berwick Railway No 156, the other *Jenny Lind*, was built by Coulthard & Co in September 1847. *(J S MacLean)*

Fig 1.18
York, Newcastle & Berwick Railway No 177, built by R & W Hawthorn in July 1848 and withdrawn in 1872. *(J S MacLean)*

Plate 1.1
YN&BR. A monochrome rendering of a coloured drawing of No 177 made by driver W G Brown during the 1860s. The livery represented will be Saxony green with Brunswick green borders; black panel bands and boiler lagging, edged with a yellow line; vermilion buffer beams; black smokebox and running plate. *(J S MacLean collection.)*

Plate 1.2
York, Newcastle & Berwick Railway. No 207, built by R & W Hawthorn in 1849, piloted the Royal Train between Newcastle and Berwick when Queen Victoria opened the High Level Bridge over the Tyne in September 1849. It is seen here with a Fletcher chimney and dome and bearing the duplicate number 207A, painted on the cab side. It was probably transferred to the duplicate list in October 1878.

3ft 7in and 4ft 9in wheels together with cylinders 15in by 20in (Fig 1.20); its tender had 3ft 1in wheels and a wheelbase of 10ft 0in. Another small 2-4-0 built in 1851 had only 4ft 0in coupled wheels. Three 2-4-0s of 1853 had 4ft $1^{3}/_{16}$in leading and 6ft 2in coupled wheels with 15in by 22in cylinders (Fig 1.21). The driving wheels were flangeless, the firebox had a transverse water space and the boiler pressure was 100psi.

The YN&BR continued to use 0-6-0s for goods and mineral traffic, but from 1852 adopted an increased wheel diameter of 5ft 0in coupled with slightly larger cylinders: 16in by 24in; twenty-five such locomotives were built by a variety of suppliers between 1852 and November 1854 (Plate 1.3). Unusually, in 1849 the YN&BR itself built one 0-6-0, having 4ft 6in wheels.

A number of tank engines were running on the YN&BR in the early 1850s, some being rebuilds of earlier locomotives. One was a 2-4-0, built for the GNER in 1839 and rebuilt in 1852 as a 2-4-0T (Fig 1.22 and Plate 1.4). This retained its 3ft 6in and 4ft 6in wheels, 13in by 18in cylinders and outside frames; water was carried in two tanks: one beneath the bunker and another under the boiler in front of the coupled wheels.

YN&BR Locomotive Livery
The boilers of the early locomotives would be lagged with painted (or stained) wood battens held in place by metal hoops, but from the late

Fig 1.19
Nos 185-9, supplied by R Stephenson & Co were among the few domeless locomotives on the York, Newcastle & Berwick Railway. *(J S MacLean)*

Fig 1.20
York, Newcastle & Berwick Railway No 63, built by Robert Stephenson & Co in 1849. It had T-shaped spokes on the leading wheels and tender wheels of a type usually found on rolling stock. The locomotive wheelbase was 11ft 6in equally divided. *(J S MacLean)*

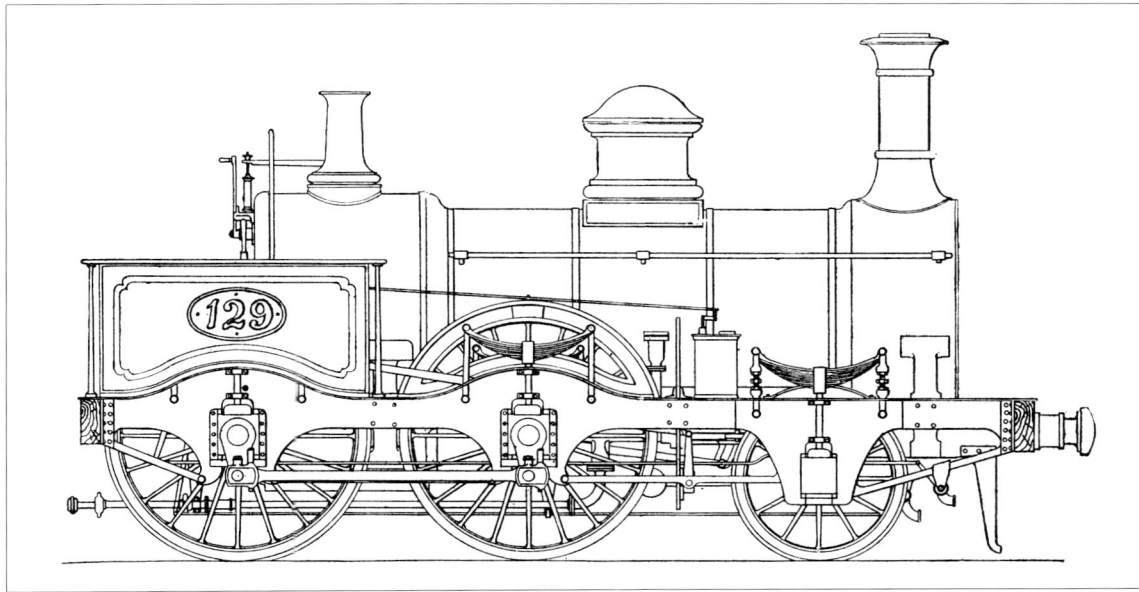

Fig 1.21
York, Newcastle & Berwick Railway No 129, though supplied by Stephenson's in 1853, shows *Jenny Lind* influence in the ornamental safety valve and dome cover. *(J S MacLean)*

Fig 1.22
York, Newcastle & Berwick Railway No 73 had been built for the GNER by R & W Hawthorn as a 2-4-0 in September 1839 and was also rebuilt by them as this 2-4-0T in 1852. *(J S MacLean)*

Plate 1.3
North Eastern Railway No 231 was built by E B Wilson & Co in June 1854, their works number 387 showing on the works plate fixed below the dome. It has received a basic cab, contrived above the original side-sheets, and a Fletcher chimney.
(J F Mallon collection)

1840s it was customary for the battens to be covered by sheet-iron cleading. One of the first new locomotives acquired by the YN&BR - Coulthard's *Jenny Lind*, built in September 1847 - had exposed wood lagging described as being "stained purple brown". (Fig 1.17)

A retired railwayman, George M Watson, who had been a booking clerk at the N&BR terminus in Newcastle and afterwards at the new Central Station from 1850, stated that the livery of the N&BR/YN&BR locomotives had been green. Incidentally, he also recalled that the guards wore green uniforms.[5]

There is also a reference to the locomotive livery, giving the actual shades of green, in a YN&BR minute of 1853: "Body Saxony or Quakers green. Brunswick green borders." YN&BR numberplates were rectangular, with concave corners and polished brass borders and numerals, the background being painted red; they were usually affixed to the footplate side-sheets, although the 2-4-0s Nos 185-9 originally had their numberplates on the sides of the high-topped 'haystack' firebox. Replicas of the numberplates were painted on the sides of the tenders. (Colour Plates 1 & 2)

---

[5] Letter from G M Watson to J S MacLean

Plate 1.4
North Eastern Railway No 73, rebuilt from a GNER 2-4-0, seen at Alnwick in the 1870s. Driver Douglas is standing in the doorway with Fireman Weddell sitting on the bunker. *(J F Mallon collection)*

The liveries of several former YN&BR locomotives were recorded by W G Brown in some coloured drawings which he made in the 1860s. (Brown joined the NER in 1863, became a fireman in 1866, a driver in 1869 and was night foreman at York sheds when he retired in 1912). His drawing of 2-4-0 No 144 (Colour plate 1) shows the locomotive as he knew it; although the shades of green on the painting may not be exactly right it is certain that they represent Saxony green as the main colour, Brunswick green for the borders and light reddish brown for the inside frames. The 2-4-0 No 177 was drawn a few years earlier; it is not known if the drawing has survived, but it is evident from the photograph (Plate 1.1) that the livery was similar to that on No 144, except that the outside frames of the locomotive were painted Saxony green to match those of the tender.

Decorated locomotives were occasionally provided for special events. For the official opening of Newcastle Central Station by Queen Victoria, on 29 August 1850, the YN&BR three-cylinder 4-2-0 No 77 was described as being,
" . . tastefully painted, carefully polished and highly ornamental. On the dome was placed a crown and in front of the smokebox was a piece of fine carved work representing St George and the Dragon; above the carving was a cushion formed of crimson silk velvet, on which was placed another crown. The tender was ornamented in keeping with the engine, having the royal arms beautifully painted on each side, with the emblems representing St. George and St. Andrew" (*Gateshead Observer*).

The YN&BR contributed 209 locomotives to the combined stock of the NER. (NER Nos 1-209)

The York & North Midland Railway (Y&NMR)
The second largest constituent of the NER was the Y&NMR, which originated as a line linking York to the North Midland Railway (NMR) at Altofts, one mile north of Normanton. The first section of the Y&NMR was opened between York and Milford on 29 May 1839, a connection being made at the latter place with the already existing Leeds & Selby Railway (L&SR). The opening throughout to Altofts Junction took place on 1 July 1840 to coincide with the opening of the NMR, thereby closing the last gap in the line of rail communication between London and York.

In November 1840 the Y&NMR leased the L&SR – opened in September 1834 – and purchased that company's locomotives and rolling stock. The L&SR was vested in the Y&NMR on 23 May 1844. The L&SR locomotives all bore names but their livery is not known.

Although Tomlinson[6] states that the L&SR owned nineteen locomotives, details are known of only sixteen, all of which were 'singles'. A small 0-2-2, built in 1830 and acquired from the Liverpool & Manchester Railway in 1834, had 5ft 0in driving wheels. Two 2-2-0s were supplied in 1834 with 5ft 3½in driving wheels and a further four with 5ft 0in driving wheels. No details are known of five 2-2-0s built in 1839 by Kirtley & Co of Warrington. In 1837 two 2-2-2s had 5ft 0in driving wheels whilst two more delivered in 1839 had 5ft 6in driving wheels.[7]

Originally the L&SR locomotives were used indiscriminately for both passenger and goods work. However, all seven of those delivered in 1839 were for passenger trains, and all the older ones were then

---

6   Tomlinson *op cit*

7   Francis Whishaw, *The Railways of Great Britain and Ireland*, 1839, states that the two 1837 locomotives were four-wheeled, but the *Leeds Intelligencer* of 2 December 1837 states that one of the locomotives delivered in that year had four wheels of 5ft 0in diameter and two wheels of 3ft 0in diameter - hence either 2-4-0 or 0-4-2.

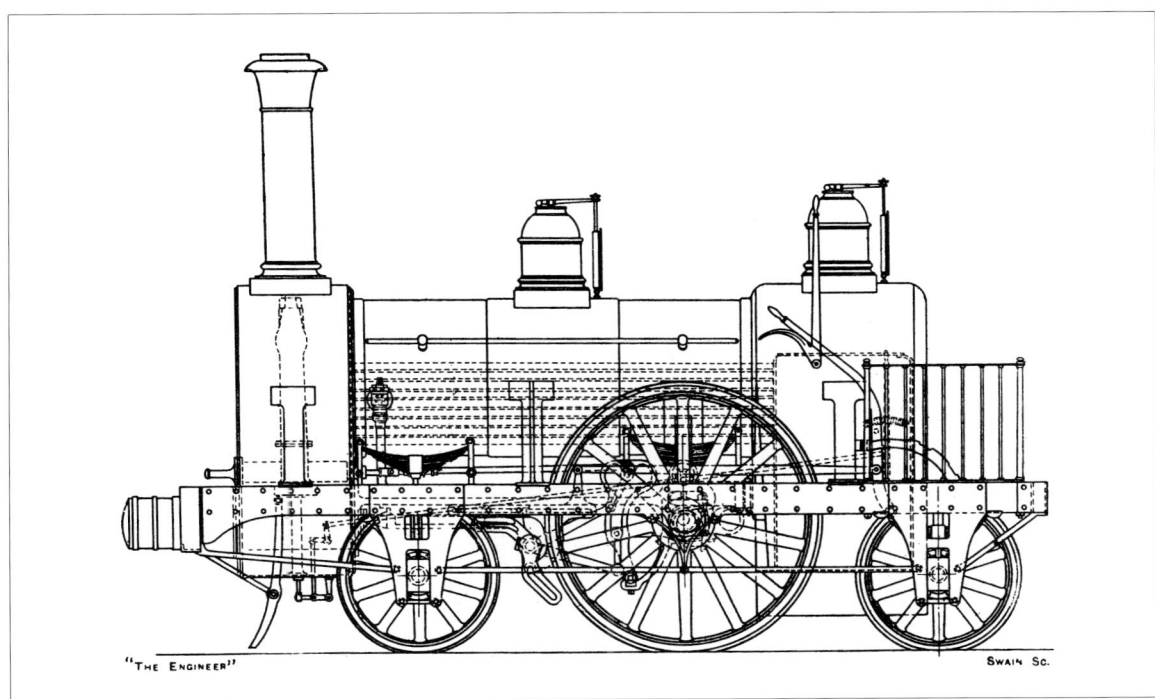

Fig 1.23
Hull & Selby Railway 2-2-2 built for John Gray by Shepherd & Todd in 1840, with Gray's 'horse-leg' motion. These locomotives had cylinders 12in or 13in by 24in.
(*John Swain*)

transferred to goods working. At the time of Whishaw's visit there were only five tenders, and five new ones were being built on four wheels to carry 50 cubic feet of coke and 700 gallons of water.

Three more companies were amalgamated with the Y&NMR; one of these, the Whitby & Pickering Railway (W&PR), which was taken over on 30 June 1845, had no locomotives, being worked by horses throughout – except for rope haulage on the inclined plane at Goathland, initially self-acting and later worked by a stationary engine, which continued until after incorporation in the NER. The other two were the Hull & Selby Railway (H&SR) and the East & West Yorkshire Junction Railway (E&WYJR).

The Hull & Selby Railway was acquired by the Y&NMR on 1 July 1845, exactly five years after its opening. The earliest locomotives were six 2-2-2s built by Fenton Murray & Jackson in 1840 and having 5ft 6in driving wheels. In 1842 they supplied a more powerful one with 6ft 0in driving wheels.

The H&SR Locomotive Superintendent, John Gray, was one of the earliest to design his own locomotives. In 1838 he patented the first form of expansion valve gear used on locomotives, which became known as the 'horse-leg' motion. A notched sector plate was provided for the reversing lever so that the cut-off could be held in a selected position, but the weight of the valve gear necessitated a supplementary lever which brought the reversing lever within reach and strength of the driver's arm. The very long-travel valves, about six inches in full gear, had 1½in lap and $^3/_8$in lead. A dome was placed on the firebox and another in the usual position on the boiler, each having a vertical steam-collecting branch pipe connecting with the main horizontal steam pipe, the regulator being at the end of this pipe in the smokebox. Six 2-2-2s were built to this arrangement by Shepherd and Todd in 1840 (Fig 1.23) with 6ft 0in driving wheels, 3ft 6in carrying wheels and a pressure of 90psi. Mixed framing was used, with bearings in the outside frames for the carrying wheels and inside bearings for the driving wheels.

The company soon adopted 0-6-0s for goods and mineral traffic, four being built by Shepherd & Todd about 1842 with 5ft 6in wheels. The H&SR built two at its Hull workshops in 1844 with 5ft 0in wheels and Gray's 'horse-leg' motion. (Figs 1.24-5) The cylinders were inclined upwards, and boiler pressure in the later ones was 100psi. Only the leading wheels had outside bearings, and the coupling rods worked on pins in the form of cranked shafts between the frames and wheels. The leading springs were thereby placed as widely apart as possible to prevent the locomotives rolling. By 1844, perhaps earlier, balance weights had been placed on the wheel rims. In the spring of 1854 one of the four earlier 0-6-0s was fitted with patent rotary valves which were soon replaced by ordinary flat valves. All those locomotives with Gray's valve gear were driven on the left-hand side.

No record of the H&SR locomotive livery has been found, although it is known that the boilers were lagged with exposed wood battens which may have been stained rather than painted. All the locomotives were named and numbered.

The East & West Yorkshire Junction Railway (E&WYJR) was taken over on 1 July 1851. The line had been worked originally by the YN&BR and latterly by E B Wilson & Co, of Leeds. Only one E&WYJR locomotive was taken into Y&NMR stock, this being a small Crampton-type 0-4-0T with an intermediate crankshaft. It became NER No 273 and in its later years was one of the locomotives selected to haul an Officer's Saloon carriage; as such it is described in Chapter 10.

The first passenger locomotives supplied to the Y&NMR were of the 2-2-2 type and appear to have also hauled goods and mineral trains for the first few years. The earliest were supplied by Stephenson's in 1838-9 and had 3ft 9in carrying wheels and 5ft 6in driving wheels; a further batch, built in 1839-40, varied only in having slightly larger boiler and cylinder diameters. In 1840 Shepherd & Todd supplied two 6ft 0in 2-2-2s with Gray's expansion valve gear, identical with those then being supplied to the H&SR (see Fig 1.23).[8]

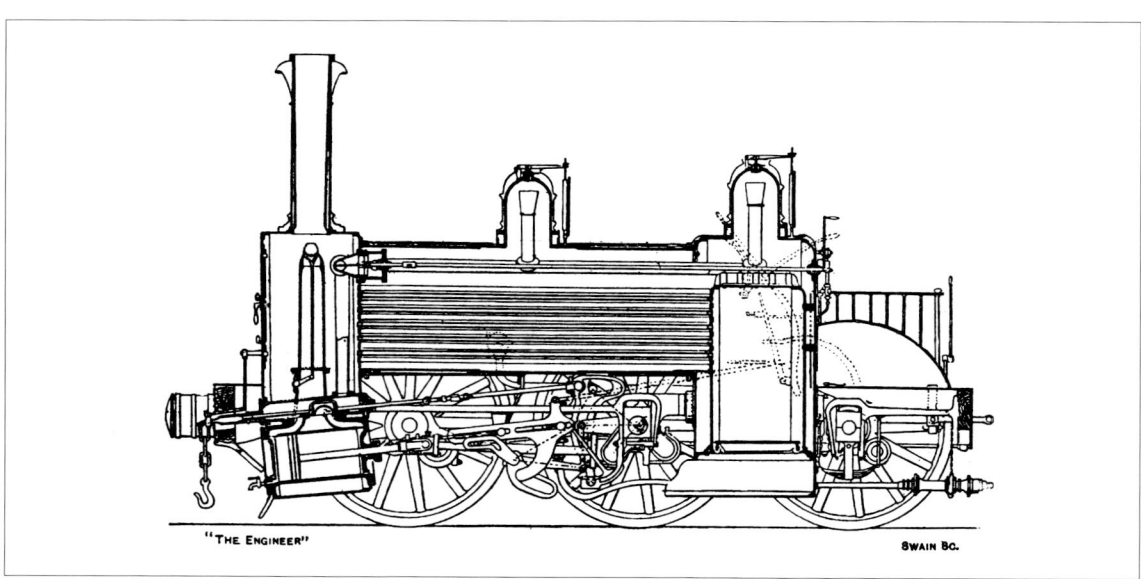

Fig 1.24
Hull & Selby Railway 0-6-0 showing Gray's 'horse-leg' motion. *(John Swain)*

Fig 1.25
Hull & Selby Railway. Exterior view of the 0-6-0 portrayed in Fig 1.24. It had cylinders 16in by 24in. *(John Swain)*

Engineers had long known that heat from the firebox passed through the tubes and chimney too quickly and was wasted. Robert Stephenson experimented by increasing the length of the boiler and tubes to utilise more of the heat and patented his idea, also placing the valves in a steam-chest between the cylinders. The 2-2-2 *Prince of Wales* (Plate 1.5), built in 1841, was one of the first locomotives with this arrangement and had 3ft 0in carrying wheels, 5ft 6in driving wheels and 14in by 20in cylinders. Stephenson's supplied four other 2-2-2s that year, two with 14in by 18in cylinders and two with 13in by 22in cylinders. In 1844 two 2-2-2s were acquired, which may have been built in 1841 and had 5ft 0in driving wheels, whilst a further three, built in 1847 and 1851, had 5ft 6in driving wheels.

When the H&SR was taken over in 1845, John Gray became Locomotive Superintendent to the London & Brighton Railway, taking delivery two years later of the first of E B Wilson & Co's celebrated *Jenny Lind* 2-2-2s (Fig 1.26), which gave its name to the rest.[9] These were designed by Wilson's chief draughtsman David Joy, whose valve gear was to be adopted on the NER by T W Worsdell in the 1880s. During 1847-8 eleven were supplied to the Y&NMR, becoming Nos 88, 90-96, 103-4 and 114. They had 4ft 0in carrying wheels, 6ft 0in driving wheels (in one case 5ft 6in), 15in by 22in cylinders and a pressure of 120psi. The original *Jenny Lind* had a safety valve mounted over the firebox, whereas those supplied to the Y&NMR had spring-balance safety-valves on top of the dome and only a square cover over the firebox manhole. A photograph taken in 1876 shows NER No 319 – formerly Y&NMR No 88 – with a lengthened wheelbase but still retaining the original boiler with its fluted "Wilson" dome etc (Plate 1.6).

The 2-4-0 type was introduced to the Y&NMR in 1842 with a locomotive having 5ft 6in coupled wheels. Seven more, built by Stephenson's in 1845-8, also had 5ft 6in coupled wheels, and one supplied by Longridge in 1846 had 6ft 0in coupled wheels. Six long-boiler 2-4-0s, ordered by the Locomotive Superintendent, Thomas Cabry, were delivered in 1846 with 6ft 0in coupled wheels (later

---

8   J S MacLean used the same drawing (Fig 1.23) to illustrate the former as E L Ahrons did the latter.

9   Jenny Lind, the *Swedish Nightingale* was enjoying fame as an opera singer.

Plate 1.5
York & North Midland Railway. Stephenson's long-boiler *Prince of Wales*, built in October 1841 and portrayed in a painting made two years later. The livery is thought to comprise varnished or green boiler lagging; green locomotive wheels, splashers etc and tender tank, picked out with black and white lines; tender frames possibly chocolate colour; red buffer beams. Renumbered 261 by the NER, it was withdrawn about 1868.
(Photograph - J M Fleming collection, but present location of painting unknown.)

Plate 1.6
York & North Midland Railway No 88 was the first of E B Wilson's *Jenny Linds* to be supplied to the Company and actually bore that name. Built in 1847, it is seen in virtually original condition as NER No 319 on the Locomotive Stores train at Scarborough in 1876. Its number was taken by a new BTP in June 1879.

(Ken Hoole collection)

Fig 1.26
An E B Wilson & Co *Jenny Lind*, as supplied in 1848 to the Midland Railway.

(John Swain)

Fig 1.27
York & North Midland Railway 2-4-0 supplied by E B Wilson & Co in 1846. There is some doubt about the number depicted for this locomotive. *(J S MacLean)*

Fig 1.28
York & North Midland Railway. Stephenson long-boiler 0-6-0. *(E L Ahrons)*

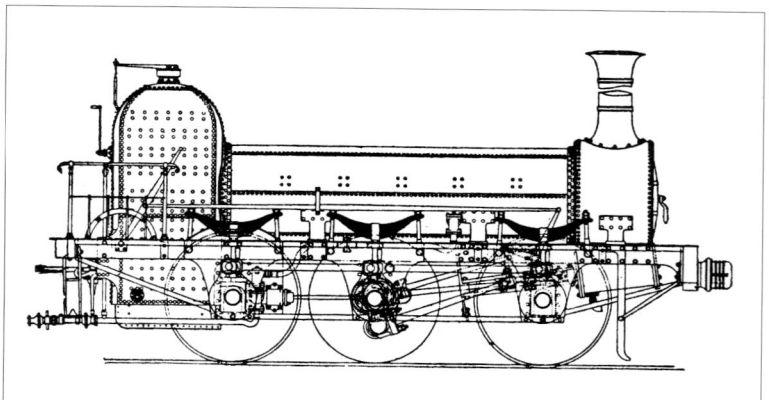

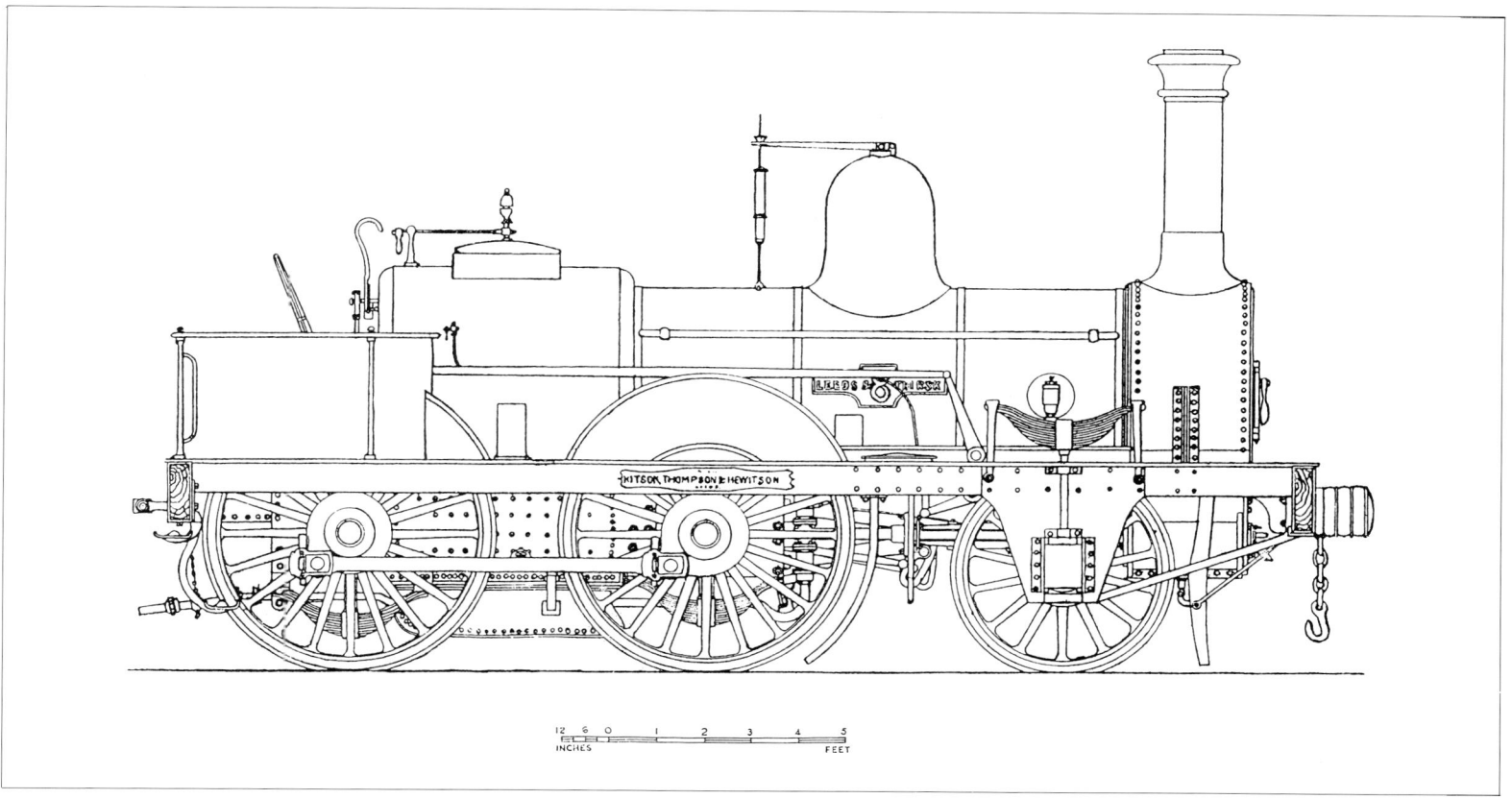

Fig 1.29
Leeds & Thirsk Railway 2-4-0 No 8, supplied by Kitson, Thompson & Hewitson in July 1849. *(J S MacLean)*

Plate 1.7
York & North Midland Railway. Another *Jenny Lind*, supplied as No 95 in November 1847 and seen as NER No 326 after receiving a new boiler in November 1871 and a Fletcher cab. *(Ken Hoole collection)*

reduced to 5ft 6in), 15in by 20in cylinders and a pressure of 90psi. They had Stephenson link motion and flangeless driving wheels (Fig 1.27). Four built in 1847 and three in 1848 had 5ft 6in coupled wheels, whilst another two built in 1847 had 6ft 0in coupled wheels and 15in by 22in cylinders. Ten 2-4-0s built in 1848-9 by Stephenson's and Kitson, Thompson & Hewitson had only 5ft 0in coupled wheels.

Two locomotives of the 0-4-2 type were supplied in 1841, one with 5ft 0in and the other with 5ft 6in driving wheels. A 4-2-0, similar to

Fig 1.30
Leeds & Thirsk Railway 0-6-0 supplied by Kitson, Thompson & Hewitson in 1848.
(John Swain)

the 'A' locomotive of the GNER (Fig 1.4), was built by Robert Stephenson & Co in 1845, followed by three in 1846 with smaller, 6ft 0in driving wheels and 15in by 22in cylinders and a further three in 1848 with cylinders of the same size but 6ft 6in driving wheels.

The Y&NMR also adopted 0-6-0s. Between 1844 and 1853, eleven were built with 4ft 6in wheels and 15in by 24in cylinders, the first two being depicted in Fig 1.28. A single 0-6-0 supplied by Longridge in 1845 with 4ft 6in wheels had been ordered by the H&SR, though it may not have had Gray's valve gear. Wheels of 5ft 0in diameter were used for six 0-6-0s built in 1847-8 by Charles Todd & Co, whilst 4ft 9in wheels were used for nine locomotives built between 1848 and 1853. Another 0-6-0 built in 1853 had 5ft 1¼in wheels.

A number of tank locomotives were acquired between 1841 and 1845. Though the details of some are unknown, a 2-2-2T of 1842 is known to have had 5ft 6in driving wheels while two 1843 locomotives, thought to be 0-6-0Ts, had 4ft 6in wheels.

The livery of the early Y&NMR locomotives is illustrated in a contemporary painting of the long-boiler 2-2-2, *Prince of Wales*, which formerly hung in the NER Institute at Gateshead (Plate 1.5). The boiler may have had stained wood lagging, the metalwork being painted green above the level of the footplate and chocolate colour below, with black and white lining.

Until August 1845 all locomotives were both named and numbered but names were not subsequently allocated to new locomotives delivered to the Y&NMR, and at about that time the locomotives taken over from the L&SR and H&SR were given numbers in the Y&NMR stock list. The Y&NMR contributed 114 locomotives to the stock of the NER. (NER Nos 245-358)

The Leeds Northern Railway (LNR)

The Leeds Northern Railway was the third constituent of the NER. It originated as the Leeds & Thirsk Railway (L&TR), the first section of which was opened on 5 January 1848, but assumed the later title on 8 August 1851 in anticipation of the opening of its extension to Stockton in 1852.

The LNR had heavier gradients than the Y&NMR or YN&BR so more powerful locomotives were required. Three 2-4-0s built by Kitson, Thompson & Hewitson in 1847 had 5ft 6in coupled wheels and 15in by 22in cylinders, whilst four built in 1848-9 had 5ft 6in coupled wheels with 16in by 20in cylinders (Plate 1.8). A further six 2-4-0s supplied by Kitson's in 1849 had 4ft 6in leading wheels, 6ft 0in coupled wheels, 16in by 22in cylinders and oval boilers 10ft 6in long and 3ft 9½in by 3ft 7½in on their cross axes; No 8 is depicted in Fig 1.29. Their plain sheet-iron domes were a Kitson feature from about 1848. Another 2-4-0 built by E B Wilson & Co in 1850 had coupled wheels and cylinders the same size as No 8, with a characteristic Wilson fluted dome.

In 1848-9 five 0-6-0s were built with 4ft 9in wheels, 17in by 24in cylinders and a boiler 11ft 0in by 3ft 8¼in (Fig 1.30). A peculiarity of both the designs shown in Figs 1.29 and 1.30 is that the locomotives are in forward gear, by an arrangement of the motion, although their reversing levers are shown in the position customary for backward gear. A further five 0-6-0s, also with 4ft 9in wheels, were delivered in 1849-50, whilst four, similar to some Stockton & Darlington Railway (S&DR) locomotives, were built by Gilkes, Wilson & Co of Middlesbrough in 1849 with 4ft 6in driving wheels; the LNR soon sold them on to the South Yorkshire Railway. Two larger 0-6-0s were built by Gilkes Wilson's in 1850 with 5ft 0in wheels, whilst two more built in 1852 had 5ft 6in wheels.

Plate 1.8
Leeds Northern Railway 2-4-0, supplied by R & W Hawthorn, seen as NER No 371, heading the first NER train into Ilkley Station on 1 August 1865. It was rebuilt with new frames and cylinders in 1871, and cut up at Darlington in December 1888.

In 1852, a 2-2-2 well tank locomotive, built by Kitson, Thompson & Hewitson in 1851 and displayed at the Great Exhibition in Hyde Park, was purchased by the company. It had 3ft 6in carrying wheels with outside bearings, 6ft 0in driving wheels and 11in by 22in cylinders (Fig 10.1). Named *Aerolite*, it had been painted blue for the exhibition. *Aerolite* was later to be the choice of the NER Locomotive Superintendent for hauling his saloon. The many fundamental changes in the design and appearance of *Aerolite* between 1851 and the present day are described in Chapter 10.

Some of the early L&TR locomotives had somewhat large numberplates incorporating the company's name, as seen in the drawing of the 2-4-0 No 8 (Fig 1.29). However, no record of the L&TR or LNR liveries has been found. The LNR handed over 29 locomotives to the NER. (NER Nos 359-387)

Joint Working and Amalgamation 1853-4

A provisional agreement between the YN&BR, the Y&NMR and the LNR for the joint working of their traffic came into effect on 1 April 1853. This was followed by amalgamation, authorised by Parliament on 31 July 1854, as a result of which the Y&NMR and LNR were dissolved and vested in the YN&BR to form the North Eastern Railway. A fourth constituent of the NER was the Malton & Driffield Junction Railway (M&DJR) which had opened on 1 June 1853 and was incorporated in the NER on 28 October 1854.

The Workshops

The headquarters and principal workshops of the NER Locomotive Department developed around the former Gateshead terminus of the N&DJR at Greene's Field which, although opened as recently as 18 June 1844, had become redundant after the opening of a new

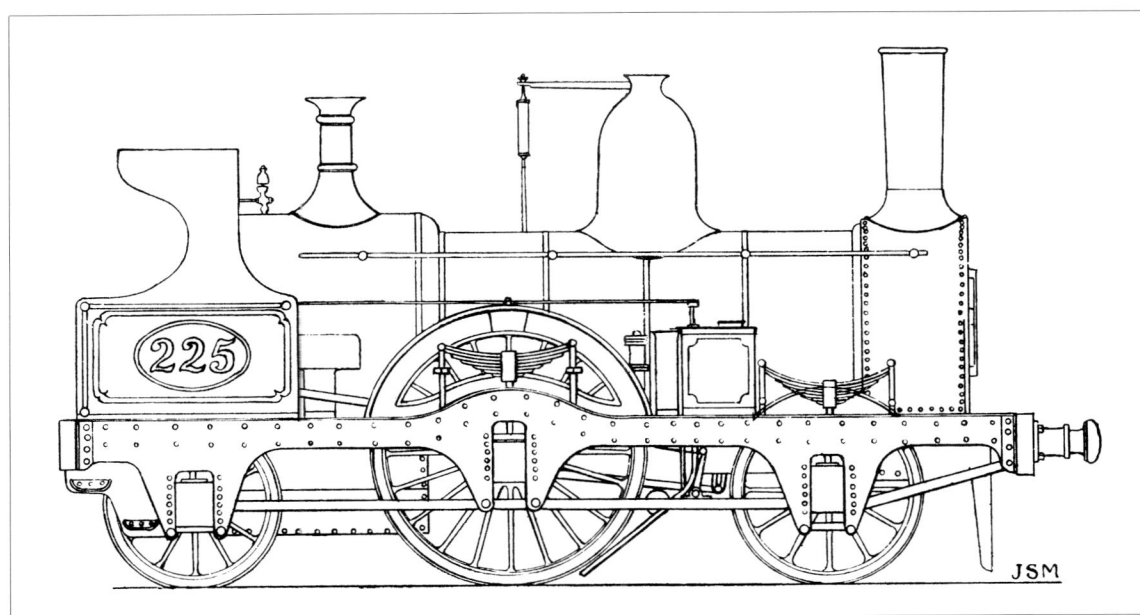

Fig 1.31
North Eastern Railway No 225, supplied by Robert Stephenson & Co in 1854, shown as reboilered and fitted with a Fletcher cab in the 1860s. *(J S MacLean)*

station at the southern approach to the High Level Bridge on 30 August 1850. The train shed and the station offices thereby became the nucleus of the "Greenesfield" Works, which replaced the former GNER workshops at Darlington, the first addition being a large workshop joined on to the northern side in 1852. Much of the handsome north facade at first survived within the new building but large openings were afterwards made in the walls, the gaps being bridged by girders to support the upper courses of masonry and the timber roof trusses. The station hotel, which still remains along with the 1852 building, became the works offices, with the locomotive superintendent housed in the former first-class refreshment room. The Greenesfield site was quite restricted and, after half a century, in 1910 the construction of new locomotives was moved to the Darlington North Road Works originally built by the Stockton & Darlington Railway. Repairs were still carried out at Gateshead until 1932, when the Works closed, only to re-open during World War II and remain in use until 1959.

The NER also inherited the Y&NMR workshops at York and the LNR workshops at Leeds, which continued in operation for many years after the amalgamation and retained a considerable degree of independence under their respective superintendents. York Works had been established at Queen Street in 1842-3 and its original buildings survived into the 1970s as the home of the Railway Museum, established by the LNER and opened in 1928. Large new erecting shops were constructed in the 1860s and these buildings remain, although the locomotive works closed in 1905.

The LNR works at Holbeck were completed in the early months of 1850 and comprised a handsome roundhouse and a long workshop range flanking the Leeds & Liverpool Canal. During 1863-4 these were augmented by a very unusual building: a repair shop in the form of a tall half-roundhouse containing an overhead travelling crane. Very little new construction was ever carried out at Leeds but some repairs continued there well after the opening of new engine sheds at Neville Hill in 1904. The buildings fell out of railway use during the 1920s and, as a result, have survived and recently been thoroughly restored.

New Locomotives 1853

Edward Fletcher, from the YN&BR, was appointed Locomotive Superintendent of the combined companies and remained in office until 1882. On 27 May 1853, less than two months after the start of the joint working arrangement, the first orders were placed for more locomotives, nine being required for new branches of the YN&B section and three for the M&DJ offshoot of the Y&NM section. Six 2-2-2s were ordered from Robert Stephenson & Co (Nos 220-5, Fig 1.31), and delivered in 1854 with 3ft 8in carrying wheels, 6ft 6in driving wheels and 15in by 20in cylinders. E B Wilson & Co delivered a *Jenny Lind* 2-2-2 from stock in October 1853 with 4ft 0in carrying wheels, 6ft 3in driving wheels and 15in by 20in cylinders (Fig 1.32). Six 2-4-0s were also ordered from Wilson's, four having 6ft 0in coupled wheels (Plate 1.9, Fig 1.33), and two (Nos 215 and 218) having 6ft 6in coupled wheels and 4ft 6in carrying wheels (Fig 1.34); these two were later rebuilt with small cabs, new chimneys and safety valves on the domes (Fig 1.35). The boiler of No 218 exploded at Holbeck station, in Leeds, on 23 July 1875 (Plate 1.10). A small 2-4-0, No 219, was supplied from Wilson's stock in December 1853 and had 5ft 9in coupled wheels.

Kitson & Co built two 0-6-0 goods locomotives with double frames for the Midland Railway, which were diverted instead to the NER (Nos 210-211) and delivered in September 1853. Four more were ordered in 1853 but built from September to November 1854 (Nos 226-229, Fig 1.36) and 15 double-framed 0-6-0s were ordered from E B Wilson & Co (Nos 230-244), and built between March 1854 and March 1855. (Fig 1.37).

A feature throughout Fletcher's period in office was the latitude given to the NER's workshops and private locomotive builders in the design

Fig 1.32
North Eastern Railway No 212, an enlarged *Jenny Lind*, supplied by E B Wilson & Co in 1853. (*J S MacLean*)

Fig 1.33
North Eastern Railway No 216, supplied by E B Wilson with 6ft coupled wheels in December 1853. *(J S MacLean)*

Fig 1.34
North Eastern Railway No 215, the larger, 6ft 6in coupled-wheel version of the locomotive shown above. *(J S MacLean)*

Fig 1.35
North Eastern Railway No 218, the counterpart to No 215, as rebuilt with a Fletcher chimney and dome and a modest cab. *(J S MacLean)*

Plate 1.9
NER No 216, supplied by E B Wilson in December 1853 but seen at Scarborough station after receiving a Fletcher boiler and cab in December 1877. It has the York version of the Fletcher cab, distinguished by circular spectacle windows, whereas the Gateshead ones were square. An additional Adams safety valve has been fitted between the chimney and the dome.

and placing of various fittings. This led to significant differences in appearance between locomotives built at the same time to the same specification and subsequently grouped in the same class.

Consolidation and Renumbering

Renumbering of the locomotives to form a single stock list took place sometime between April and September 1855. The former YN&BR locomotives – 1 to 209 – and those that had been ordered jointly in 1853 – 210 to 244 – retained their existing numbers; locomotives from the Y&NMR became Nos 245 to 358 and those from the LNR were allocated Nos 359 to 387. However Nos 380 to 383 were condemned by the NER before actually being renumbered and were replaced by four new 0-6-0s built by E B Wilson in April/May 1855, the firm accepting the LNR locomotives in part payment! Elliptical brass numberplates with raised borders and numerals were provided for the amalgamated stock: these were usually affixed to the footplate side-sheets although sometimes they were curved to fit on the sides of the boilers.

Early Rebuilds

For some years after the amalgamation, the NER continued to rely on outside manufacturers to meet the requirements for additional motive power, although most of the replacements for obsolete or worn-out locomotives were built in the railway's own workshops.

Gateshead Works dealt initially with repairs and rebuilding, including the provision of new boilers. Among the locomotives reboilered in the late 1850s was 2-4-0 No 21, its original domeless boiler with a high-topped 'haystack' firebox being replaced by a new one with a dome (Plate 1.11). No 21 had been built for the N&DJR by Robert Stephenson & Co in December 1844 and was actually the first *new*

Plate 1.10
North Eastern Railway No 218, seen outside the repair shop at Leeds Holbeck after its boiler explosion in 1875. *(J F Mallon collection)*

Plate 1.11
NER No 21, built for the Newcastle & Darlington Junction Railway by Robert Stephenson & Co in December 1844, seen after being reboilered at Gateshead in the late 1850s. It was probably withdrawn about February 1869 when a new locomotive took its number. *(K L Taylor collection)*

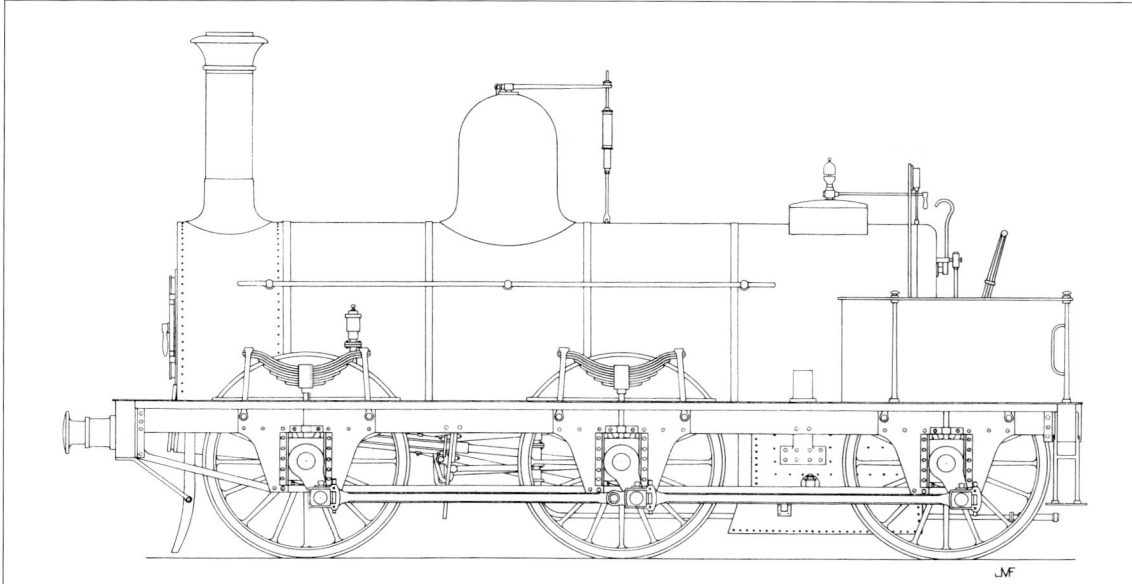

Fig 1.36
North Eastern Railway Nos 226-9, supplied by Kitson & Co in 1854. *(J M Fleming)*

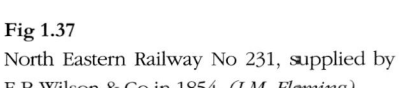

Fig 1.37
North Eastern Railway No 231, supplied by E B Wilson & Co in 1854. *(J M Fleming)*

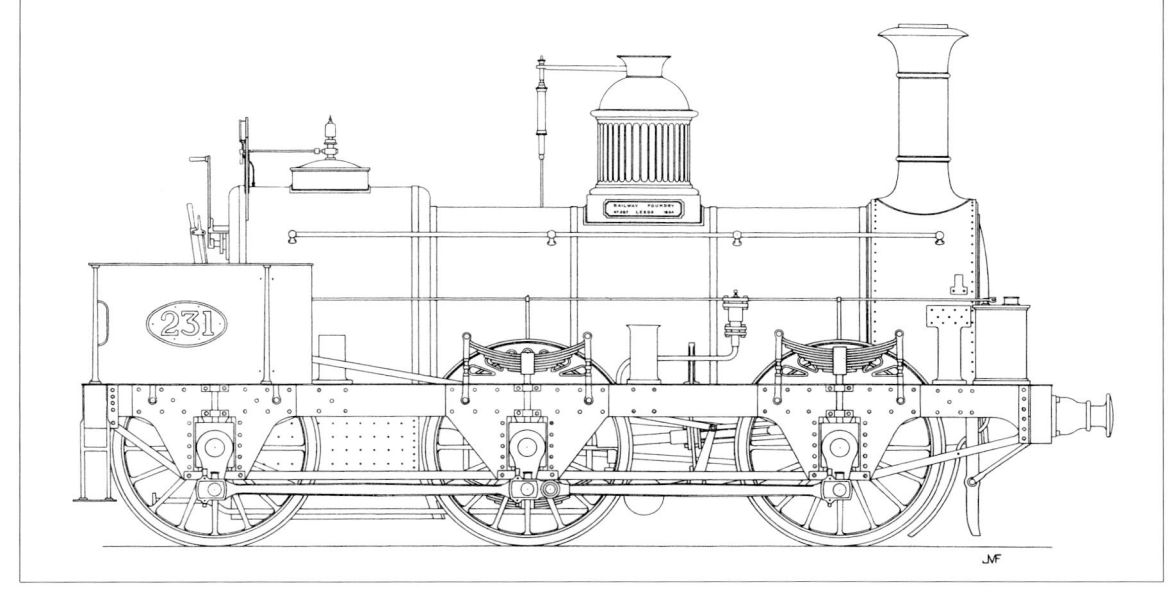

locomotive delivered to the company, the numbers 1 to 20 being allocated to acquisitions from other railways by amalgamation or purchase. Some other locomotives of the same type as No 21 were rebuilt with a longer wheelbase and compensating levers between the springs of the driving and trailing axles, after which they were said to ride very badly!

Some of the early rebuilding done at York Works produced virtually new locomotives; among these were four 2-4-0s and some small 0-4-0 tender locomotives, the latter being for passenger traffic on the Whitby Branch, on which the abrupt changes of gradient at each end of Goathland Incline precluded there being more than four wheels (Fig 1.38 and Plate 1.12). The curious arrangement of the springs at the sides of the low-pitched boiler of the 0-4-0s was also adopted for an 0-6-0 No 283 and possibly for some other locomotives.

New Locomotives 1854-1863

Most of the early locomotives built to Edward Fletcher's designs for the NER had double frames and outside cranks, the outside frames usually being of sandwich form and having separate hornplates bolted or riveted to each side of strong longitudinal beams; 0-6-0s usually had either 12in by 1in iron beams or 10in by 4in oak beams sandwiched between ½in iron flitch plates.

External details of locomotives built at Gateshead Works up to 1862 included wrought-iron chimneys capped with a bell-mouthed casting, tall domes with the Salter safety-valve levers protruding below the top flange and, usually, ornamental firebox manhole covers somewhat resembling Victorian soup tureens. By the end of 1862 the height of the domes had been reduced so that the safety-valve levers were above the top flange. Tapering 'stovepipe' chimneys with only a

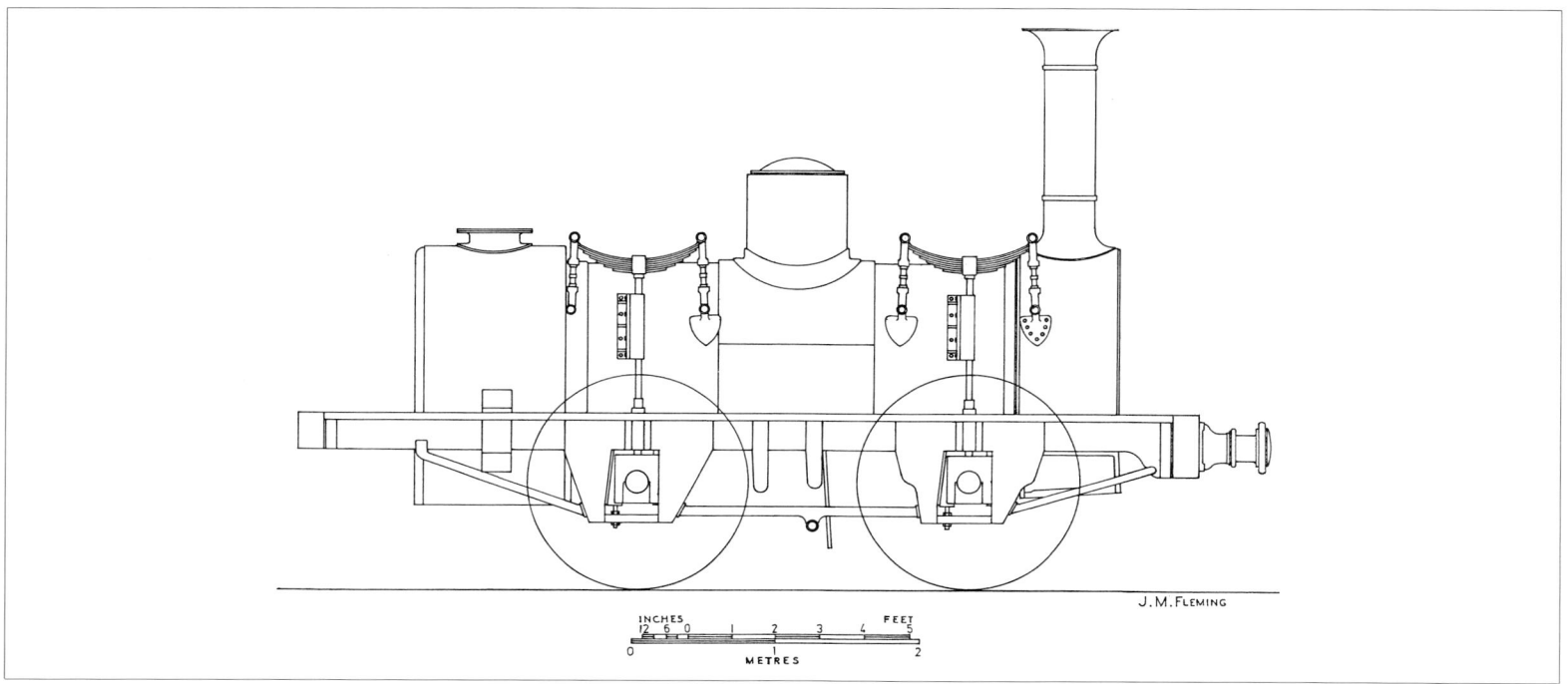

Fig 1.38
No 272. Whitby Branch 0-4-0. *(J M Fleming)*

narrow beading around the top were introduced in 1864 and gradually replaced the earlier design.

Goods and Mineral Locomotives

Goods and mineral locomotives always outnumbered those built for the NER passenger traffic, the most numerous among the 0-6-0s with outside frames being those with an equally-divided 15ft 4in wheelbase. At least 93 of these locomotives were built between 1854 and 1866 by seven different outside manufacturers, each being allowed considerable freedom in the matters of appearance and details. The ornate, fluted domes of those built by E B Wilson & Co in 1854-7

Plate 1.12
NER No 272, one of Fletcher's Whitby Branch 0-4-0s, with its correspondingly short four-wheeled tender, waiting at the north end of Pickering station. Built at York Works in December 1860, with 5ft wheels and 15in by 20in cylinders, it originally had the springs well up the boiler side, as seen in Figure 1.38. It went on the duplicate list as No 272A and was still working in June 1888, but was scrapped soon after in August.

Plate 1.13
One of the 0-6-0s supplied by Manning, Wardle in 1861 with inside frames.

Plate 1.14
Also supplied by Manning, Wardle, in 1861, No 434 has outside frames and safety valves mounted on the dome, in line with Fletcher practice.
(J F Mallon collection)

also appeared on four locomotives built in 1861 by that firm's successors, Manning, Wardle & Co (Plates 1.13-14).

The NER started to build locomotives at Gateshead Works in 1857, the first one being the outside-framed 0-6-0 No 13 which was completed in June of that year (Fig 1.39). A similar locomotive, No 133, built at Gateshead in April 1861, is shown as it appeared about 1870 after being reboilered and provided with a cab in place of its original weatherboard (Plate 1.15). Gateshead Works continued to build goods locomotives to the same basic design until 1871, various detail modifications being introduced from time to time. A few were built in 1870-1 with a 16ft 0in wheelbase, divided 7ft 8in + 8ft 4in. All, however, had 5ft 0in driving wheels and 16in by 24in cylinders. (Under the classification introduced in 1886 all these 0-6-0s were designated Class 13.)

Gateshead Works also produced a long-boiler version of the 0-6-0s described above, the first example being No 93, completed in January 1860 (Fig 1.40). In addition to those built at Gateshead, a further 75 were supplied by four different manufacturers between 1860 and 1868, most of which had sandwich-type outside frames with oak beams; all had a 'short-coupled' 12ft 4½in wheelbase, divided 7ft 1in + 5ft 3½in. It is interesting to see that No 160, built at Gateshead in April 1862, incorporated a set of second-hand wheels of an out-of-date pattern, the sixteen spokes being of circular section with T-shaped heads forged or fire-welded on and riveted to the wheel rims (Plate 1.16), and, indeed, both 4ft 0in and 5ft 0in wheels were to be found among these locomotives. The 'stovepipe' chimney and cab

Fig 1.39
No 13, (later Class 13), the first locomotive to be completed at Gateshead.
(J M Fleming/traced, J F Addyman)

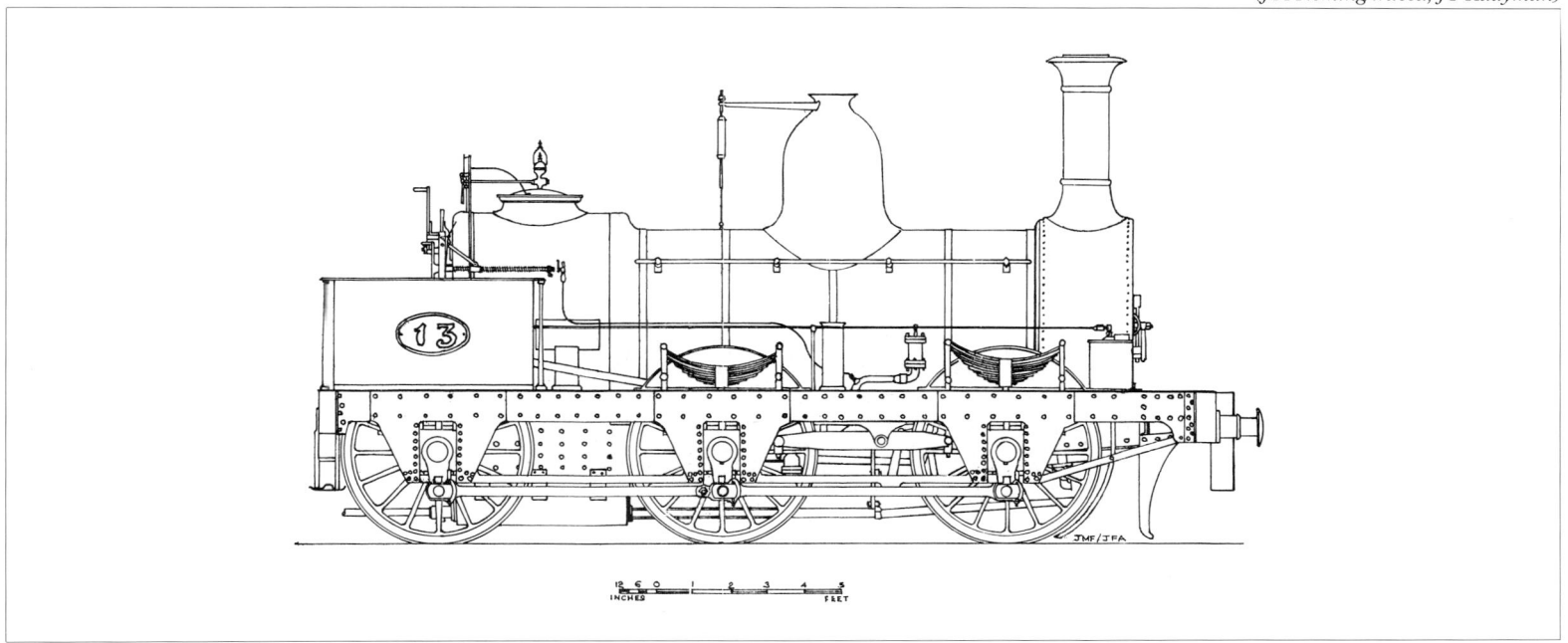

Plate 1.15
Class 13. No 133 after being reboilered and provided with a cab.
(M R Grocock collection)

were later additions, the cab being a product of the Hartlepool Works about 1870. (The 1886 designation was Class 93.)

In 1860 Gateshead Works began to add to the pre-1854 stock of long-boiler 0-6-0s for the mineral traffic. These had inside frames, a short wheelbase, 15in by 24in cylinders and wheels of various diameters between 4ft 6in and 5ft 0in (Fig 1.41). An early example – No 158, built in September 1864 – is illustrated by Plate 1.17. (The 1886 designation was Class 41, and other 0-6-0s of similar dimensions built around this time brought its total to 87.)

Passenger Locomotives

One 2-2-2 passenger locomotive with outside sandwich frames and 6ft 6in driving wheels was built at Gateshead Works in 1859 and another in 1861, the design being based on that of the 2-2-2s ordered from R Stephenson & Co in 1853. They were followed in 1861 by six larger, outside-framed 2-2-2s with fireboxes designed for burning coal instead of coke. These had 4ft 6in leading wheels, 6ft 6in driving wheels and 4ft 0in trailing wheels, with 16in by 22in cylinders. Nos 447-449, built by R Stephenson & Co, had fireboxes with a longitudinal 'midfeather' (a vertical division) (Fig 1.42), while the fireboxes of Nos 450-452, built by R &W Hawthorn, had combustion chambers extending 18 inches into the boiler barrels.

Plate 1.16
North Eastern Railway No 160, built at Gateshead in 1863 and seen with a later stovepipe chimney and Hartlepool Works cab. *(Ken Hoole collection)*

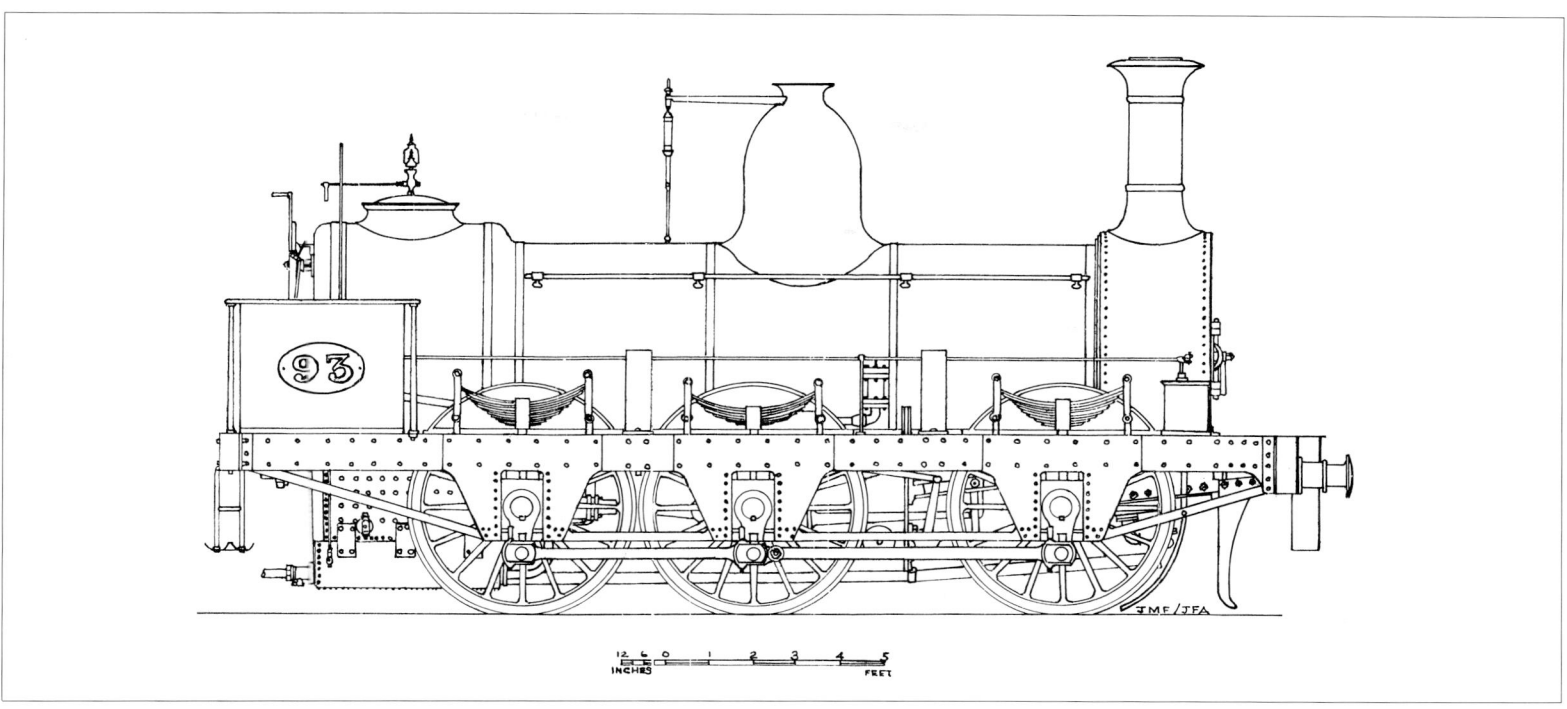

Fig 1.40
Class 93 No 93, built at Gateshead in 1860. *(J M Fleming/traced, J F Addyman)*

Each of these six locomotives had an 'exhaust cock' below the cylinders which enabled the drivers to divert some of the exhaust steam into a pipe leading to the front of the ashpan; thence the steam would be directed into the firebox through 'air induction steam jets' to accelerate the air flow over the fuel and ensure more complete combustion and less black smoke. (The Great North of Scotland Railway had adopted a similar system in 1859 using live steam and acting as a blower when the locomotive was stationary.) The air induction jets were soon abandoned, but exhaust cocks continued to be standard fittings for most of Fletcher's locomotives. Steam discharged via the exhaust cocks was passed directly into the atmosphere thereby softening the blast and giving an easy exhaust with less throwing out of unburnt fuel. Three 2-2-2s similar to the above six were built at Gateshead in 1862, their numbers being 161, 162 and 280. No 162 was rebuilt as a 2-4-0 in 1879.

Hawthorn-built No 451 was photographed about 1880 when it still retained its original boiler (Plate 1.18); the safety valve over the firebox was a later addition, as were the cast-iron chimney, steam brake gear and cab. Seen at the top of the driving wheel splasher is a transfer picture of a bunch of roses - one of a variety of unofficial decorations added by drivers during the Fletcher regime. (The 1886 designation of the 2-2-2s was Class 447, the 2-4-0 rebuild being Class 162).

The NER was one of the first among the larger English railways to cease building locomotives with single driving wheels and to construct only four-coupled locomotives for its main line passenger trains.

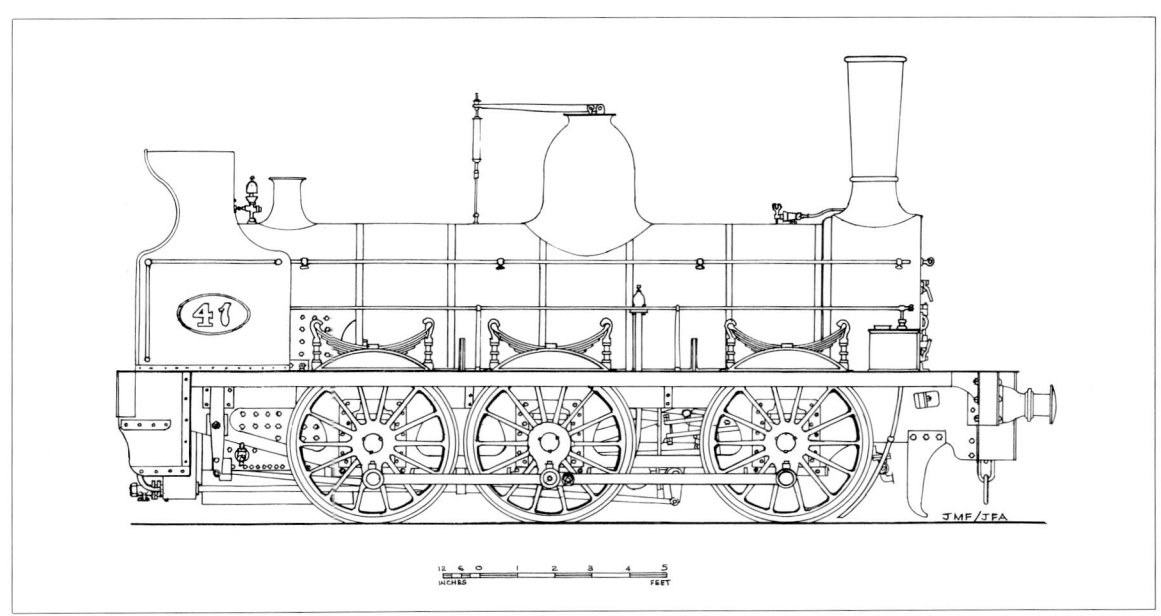

Fig 1.41
Class 41 No 41, built at Gateshead in October 1861, shown as reboilered and fitted with a cab.

(J M Fleming/traced, J F Addyman)

Plate 1.17
Class 41. No 158 built at Gateshead in September 1864.

In 1862 Gateshead Works started to build a new class of 2-4-0s with outside sandwich frames and 6ft 6in diameter coupled wheels. The first two locomotives – Nos 157 and 312 (Plate 1.19) – each had a longitudinal midfeather extending to the roof of the firebox and thus requiring two fireholes; some of the later locomotives had midfeathers extending only halfway up inside the fireboxes. No 157 initially had an old set of undersized wheels with supplementary rims to increase their diameter to fit 6ft 6in diameter tyres. Thirteen of these 2-4-0s were built between October 1862 and March 1867, followed after an interval by one more in October 1869. A similar class of 2-4-0s with smaller wheels was introduced in 1867; these had 6ft 0in coupled wheels and 4ft 1in carrying wheels, while both classes had 16in by 22in cylinders. (All were included in Class 25 after 1886.)

Fig 1.42
Class 447 No 447, built by R Stephenson & Co in July 1861. *(J S MacLean)*

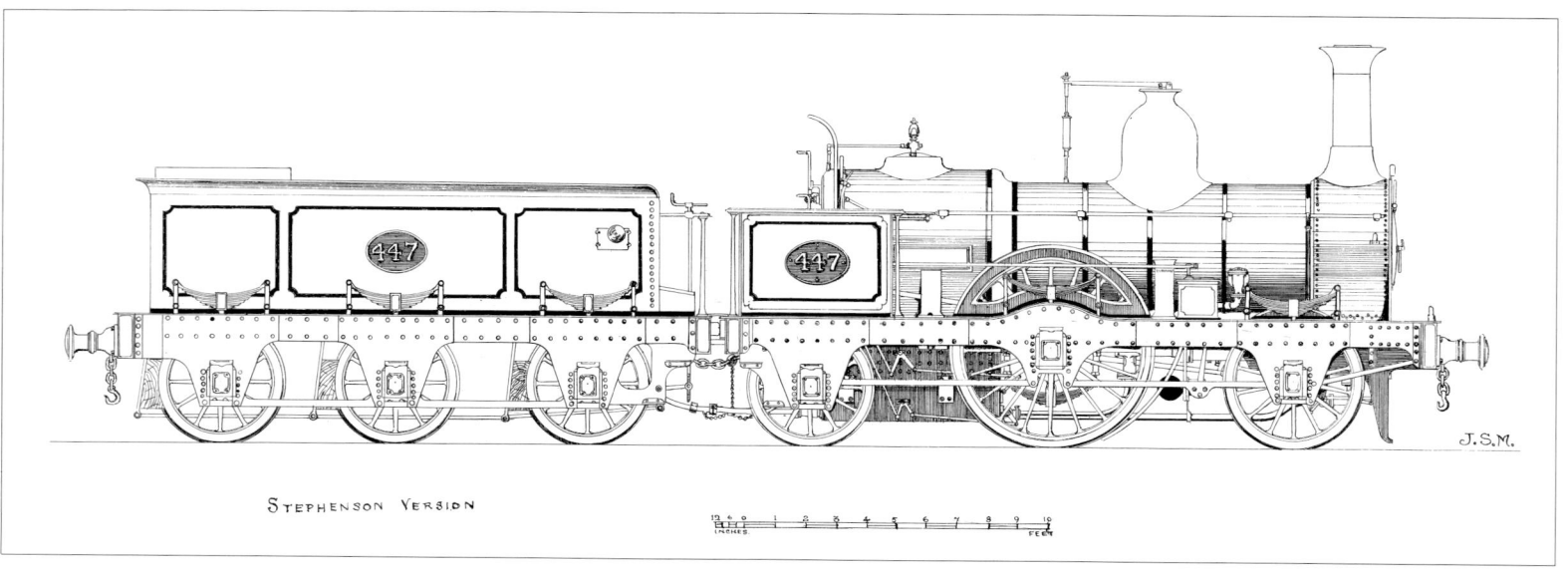

Communication Systems

In March 1856 it had been decided that bells should be fitted on all passenger locomotives to enable the guard to signal to the enginemen, the bells being operated by cords passing along the carriages below the doors. One of these bells may be seen in the illustration of 2-4-0 No 312 (Plate 1.19). Later experiments with cord communication led to an improved alarm system devised by T E Harrison, Engineer-in-Chief of the NER, being adopted in 1869. Plate 1.19 also shows Harrison's alarm equipment fitted to a locomotive without a cab; the stanchion mounted on the right-hand footplate side-sheet carried one of the bearings for a horizontal shaft extending almost half-way across the weatherboard, the shaft having a small lever at the outer end to which the operating cord was attached and a longer lever at the inner end which struck the handle of the right-hand whistle when the cord was pulled. The alarm cords, which also activated the striking mechanism of a gong in the guards van, were provided along both sides of the carriages above the doors where they could be reached by a passenger leaning out of the window. The operative cords connected to the alarm were those on the right-hand side of the train, facing the direction of travel.

Plate 1.18
Class 447 No 451, built by R & W Hawthorn in September 1861 and seen at Gateshead Works outside the former Redheugh Incline engine house. The locomotive has apparently retained its original boiler but gained a Fletcher chimney, dome and cab. The tender is one of the earliest examples of Fletcher's narrow-bodied tender with the springs above footplate level. Renumbered 1937 on 30 September 1891 and 1737 in January 1894, it was withdrawn in August 1896.

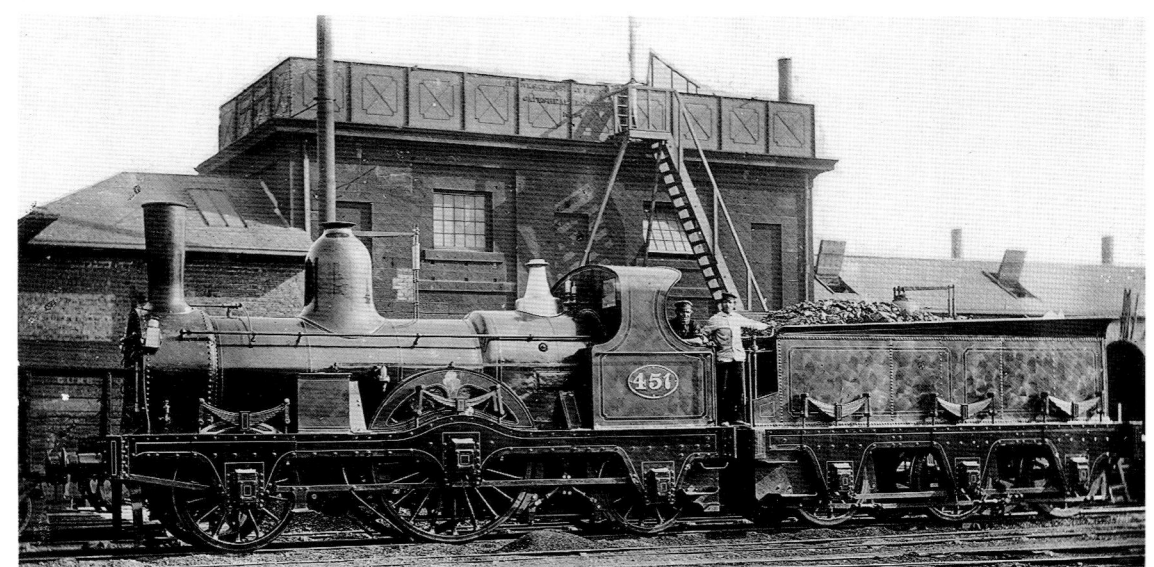

Plate 1.19
Class 25, No 312, built at Gateshead with 6ft 6in diameter coupled wheels and seen in its original condition. Note the alarm bell on the tender, superseded by T E Harrison's whistle alarm, the stanchion for which can be seen projecting up from the footplate side-sheet. The livery appears to correspond to light green with dark green borders. Renumbered 1847 in April 1890, 1780 in January 1894 and 2052 in March 1897, it was withdrawn in March 1898. The Newcastle & Carlisle Railway engine shed at the west end of Newcastle Central Station can be seen in background.

Chapter 2

Amalgamations 1862 – 1865

The Newcastle & Carlisle Railway (N&CR)
Three long-established railways, each having a sizable number of locomotives, were amalgamated with the NER in the 1860s, the first to be absorbed being the N&CR on 17 July 1862. The main line of the N&CR had been opened in several stages between March 1835 and October 1839, although Newcastle was only served by a temporary station until January 1851.

The early passenger locomotives were all 0-4-0s, those for the merchandise and mineral trains being either 0-4-2s or 0-6-0s. The 0-4-0 No 2, *Comet* (Plate 2.1), built by R &W Hawthorn & Co in 1835, was one of the first locomotives to have four fixed eccentrics for operating the valve gear. A splendid banner that was formerly exhibited in the old Museum of Science, at Newcastle upon Tyne, depicted *Comet* in colour; the boiler and firebox, dome, frames and wheels were maroon, the frames and wheel hubs being picked out with vermilion. The splasher beading, safety-valve cover, boiler cleading belts and nameplate were polished brass, the background of the nameplate being vermilion. The smokebox and chimney, cylinders and lower part of the firebox below the frames were black.

this being in addition to the nameplate on the locomotive. (There is just a slight possibility that the name panel MAY have been a lighter colour than maroon - perhaps vermilion.)

A representation of *Samson*, after being rebuilt as a 2-4-0 in 1852, occupied the centre of a banner that was reputed to have been displayed at the opening of the Alston Branch on 17 November 1852.

Plate 2.1
Newcastle & Carlisle Railway. An engraving of *Comet*, published in 1838 in F W Simms' *Public Works of Great Britain*.

Additional information about the N&CR livery is provided by a contemporary drawing of 0-4-0 No 10, *Lightning*, built in 1857, and by a primitive painting of 0-4-2 No 5, *Samson*, the latter shown as rebuilt with a "steam drier" – i.e. superheater – boiler in 1841. These two illustrations include the tenders, each showing the name of the locomotive painted in large letters within a long rectangular panel,

Although somewhat crude, the drawing included such details as pipework and the small ploughs at the front, the latter being found on all N&CR locomotives. *Samson* was one of the locomotives later fitted with a 'canopy' over the footplate, this comprising a cab with a window in each side and a wooden roof.

G M Watson – the retired Newcastle Central Station booking clerk - wrote that the N&CR locomotives in the 1850s were "... painted a bright scarlet and looked very smart with all the bright brass facings, the guards wearing scarlet gold-laced coats..."[1]. One of the early 0-6-0s, No 21, *Matthew Plummer*, built by Thompson Brothers at Wylam Ironworks, was distinguished by having a tall brass chimney, Mr. Plummer being chairman of the N&CR.

The N&CR locomotives were somewhat smaller than their contemporaries on the NER and right to the end of the former company's separate existence they had the regulator fitted on top of the firebox, the arm moving horizontally from side to side. All the locomotives had names as well as numbers.

Plate 2.2
Newcastle & Carlisle Railway. *Venus*, an 0-4-2 supplied by Thompson Brothers, of Wylam, in January 1841. It is seen here working for a contractor on the construction of the Mid-Wales Railway near Builth Wells, still apparently in original condition and bearing both the boiler nameplate and the name painted on the tender.

The 39 locomotives taken over in 1862 were allocated NER numbers 453 to 491. The names were retained for several years after the amalgamation but were eventually removed.

The Stockton & Darlington Railway (S&DR)

The most important amalgamation in the 1860s was the alliance formed by the NER with the S&DR. However, the Amalgamation Act of 13 July 1863 made provision for the former S&DR to continue to be managed by a separate Darlington Committee for the next ten years. The original line of the S&DR from the coal pits at Witton Park, north-west of Darlington, to Stockton Quay had been expanded considerably by 1863. Many of the extensions had been built by nominally independent companies (although always operated by the S&DR) by which the railway reached as far as Tebay and Penrith in the west, to Consett in the north and to Saltburn in the east.

William Bouch, who had been responsible for the S&DR locomotives since the resignation of Timothy Hackworth in 1840, remained in charge of the locomotives of the Darlington section under the new regime until his death in January 1876. Since 1849 the motive power had actually been provided under contract by the Shildon Works Co, a partnership formed by Bouch with Oswald Gilkes (Secretary of the S&DR) and Edgar Gilkes. The company also took over the S&DR workshops at Shildon and Waskerley, afterwards moving the main centre of activities to new workshops at North Road, Darlington, on 1 January 1863.

The celebrated pioneer engine, *Locomotion*, was still included in the S&DR stock at the date of the amalgamation, although it had long since been classified as "old and useless". Fortunately it had been retained as a curiosity and had been partially restored to something like its original appearance in 1857. Since June 1857 it had been exhibited on a pedestal outside Darlington North Road Station and was not finally withdrawn from stock until 1868.

Very little has been recorded about the liveries of the early locomotives. It is said that *Locomotion* was painted in gaudy colours when it hauled the inaugural train at the opening of the S&DR on 27 September 1825, but after a month or two it received a coat of dark brown pigment. A contemporary drawing of another early locomotive, 0-6-0 No 8, *Victory*, built in 1829, shows it painted black which may have been the usual 'colour' at that period.

Three small locomotives for the "coach trains" were obtained in 1830-1, but they could only be used for hauling mineral trains until after the S&DR had bought out the proprietors of the horse-drawn coaches in September 1833. The first of these locomotives was Hackworth's 0-4-0 No 9, *Globe*, built by R Stephenson & Co in 1830. In an early drawing *Globe* is shown with a black boiler and chimney, the former being almost hidden by elaborate wheel splashers painted blue and black, with green and yellow decorative work. The wheels were blue and the boiler had a copper dome in the shape of a globe.

From an early period all the locomotives had borne numbers as well as names but between 1837 and 1847 the "coach engines" were identified only by names, the numbers having been removed.

---

1 Letter from G M Watson to J S MacLean

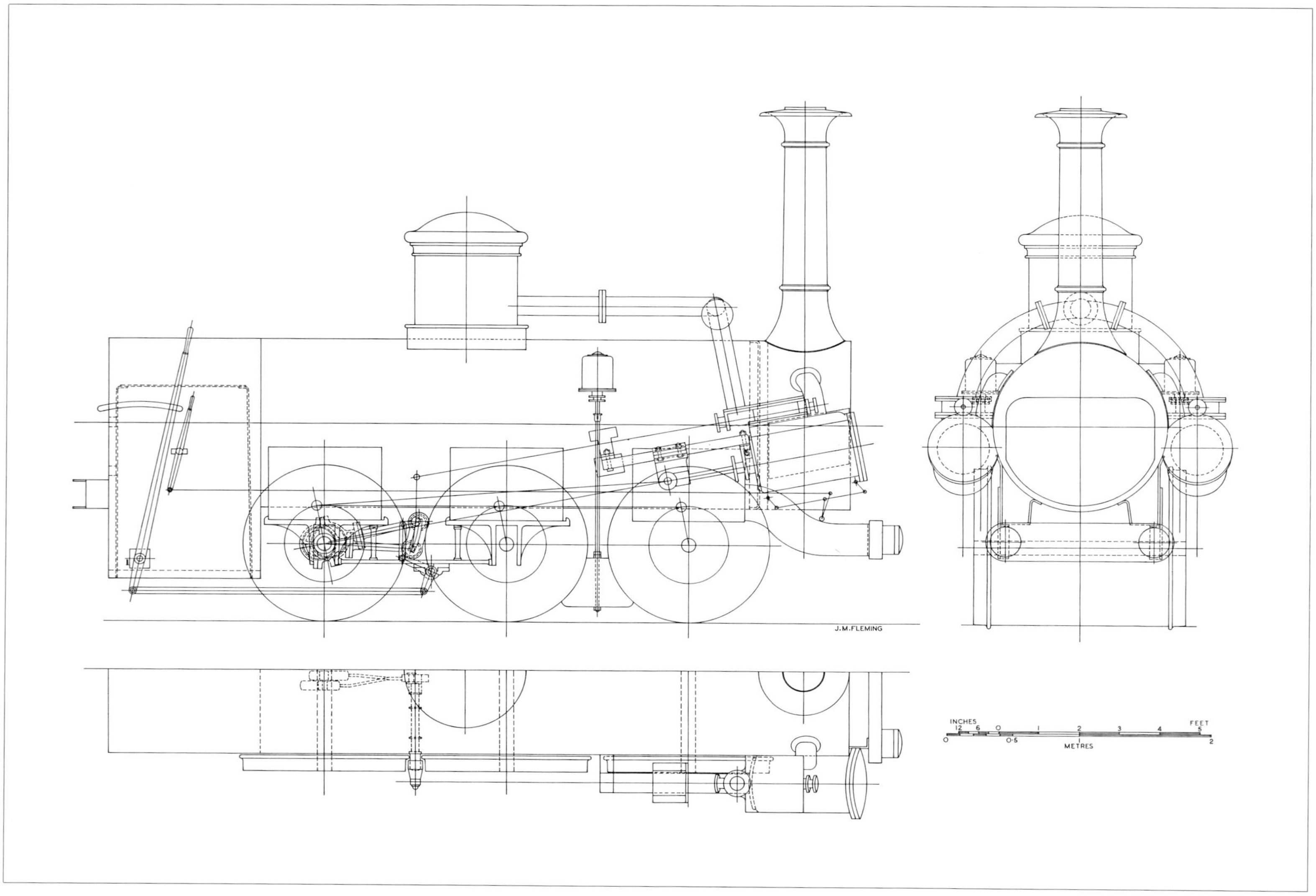

Fig 2.1
No 14 *Tees* as 'remodelled' in 1848. *(J M Fleming)*

Plate 2.3
Stockton & Darlington Railway No 23, *Wilberforce*, supplied by R & W Hawthorn in 1832 and seen outside the first roundhouse at Shildon. Essentially a Hackworth design, it had a multi-tubular return-tube boiler, hence the fireman is standing on the coal tender to the right of this view.

S&DR Mineral Locomotives 1838-1848

The various early types of mineral locomotives with vertical cylinders – some with unsprung driving wheels – were followed by Hackworth's notable 0-6-0s with inclined cylinders mounted on the side of a return-tube boiler. As in his previous designs, brackets formed from ³⁄₈in iron plates riveted to the boiler barrel served as main frames. Altogether 26 locomotives of this type were built between 1838 and 1848, successive increases being made in the size of the cylinders.

Nameplates were usually affixed to the sides of the boilers, these being either brass castings with raised lettering and borders or flat brass plates with incised letters. Numberplates were square, with either raised or incised numerals, and were to be found either on the sides of the boilers or on the domes. The numbers were also painted on the sides of the tenders. The livery was recorded in May 1837 when a Mr H Wilson, of Bishop Auckland, undertook to paint locomotives and tenders for £1 15s 0d (£1.75) each. The colours were specified as:- "Cleading green: hoops black: wheels black relieved with red. All ironwork except (i e not otherwise specified) polished black. Tenders blue and hoops (of water barrel) black: ironwork black."

The first of these new mineral locomotives was No 15, *Tory*, built in November 1838, followed by five more in 1839-40, all having boilers with a fire tube and a single return flue tube, the latter being joined to the chimney by an external elbow. They had 4ft diameter wheels and, initially, 12½in by 18in cylinders.

No 6, *Dispatch*, built in September 1839, had a multitubular return-tube boiler, the 69 return tubes being grouped around the fire tube and ending in a horseshoe-shaped smokebox which partially surrounded the firedoor. Four more similar locomotives were built in 1840-1, all having 14in by 18in cylinders.

No 7, *Prince*, built at the S&DR Shildon Works in March 1842 (Works No 1), had 15in by 20in cylinders. Three more of the same type, with 14½in by 20in cylinders, were added in 1842 by remodelling existing vertical-cylinder locomotives Nos 8, 11 and 25. Nos 8 and 25 were remodelled at Hackworth's Soho Works and were classed as 'new' when returned to the S&DR, No 25 being renumbered 12; both were given new names, becoming No 8, *Leader*, and No 12, *Trader*. (Note 'remodelling' was the word used by the S&DR for rebuilding).

All the foregoing locomotives had a tender for coal at the firegrate end and another for water at the driver's end. Large wooden water barrels were evidently provided at first but in 1845-7 these were replaced by iron tanks holding 749 gallons.

No 14, *Tees*, was remodelled from a vertical-cylinder locomotive at Shildon Works in December 1844 with new 14½in by 24in inclined cylinders. It was remodelled again in May 1848 with a firebox in place of the original firetube, 16in by 24in cylinders and a steam-operated brake (Fig 2.1). A six-wheeled tender to carry four tons of coke and 1,401 gallons of water was also provided. *Tees* was used as a banking locomotive to assist trains up the rope-worked incline between Howden-le-Wear and Crook.

No 29, *Miner*, and five more locomotives – Nos 30-34 – were built at Shildon Works between June 1845 and March 1846 with 14½in by 24in cylinders and return-tube boilers. Between May 1848 and September 1849 the boilers of all except No 31 were remodelled with fireboxes, new tubes and new smokeboxes; the position of the cylinders on the boiler barrels was unchanged but this now became the leading end. Separate tenders for coal and water respectively were retained. The return-tube boiler of No 31, *Redcar*, was never remodelled.

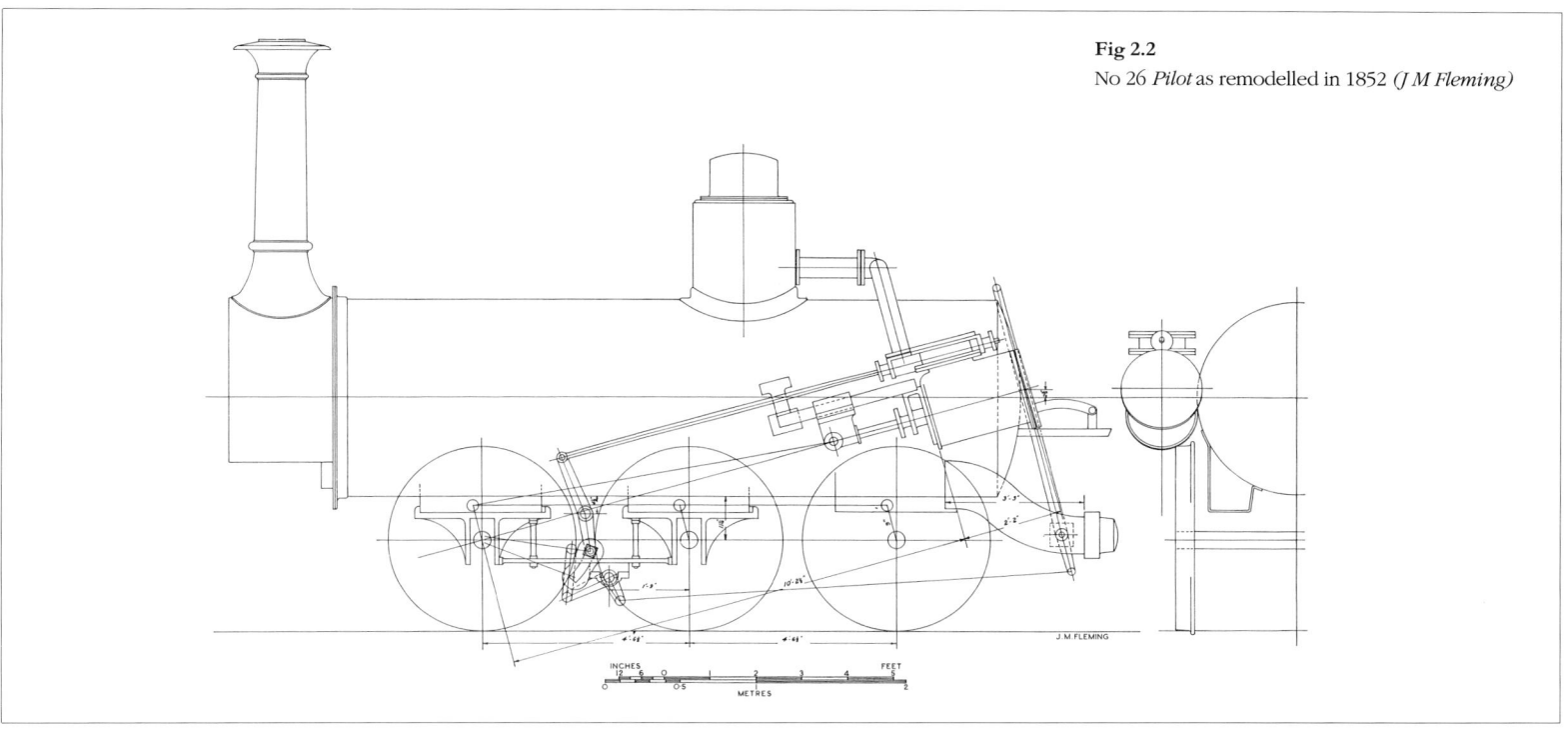

Fig 2.2
No 26 *Pilot* as remodelled in 1852 (*J M Fleming*)

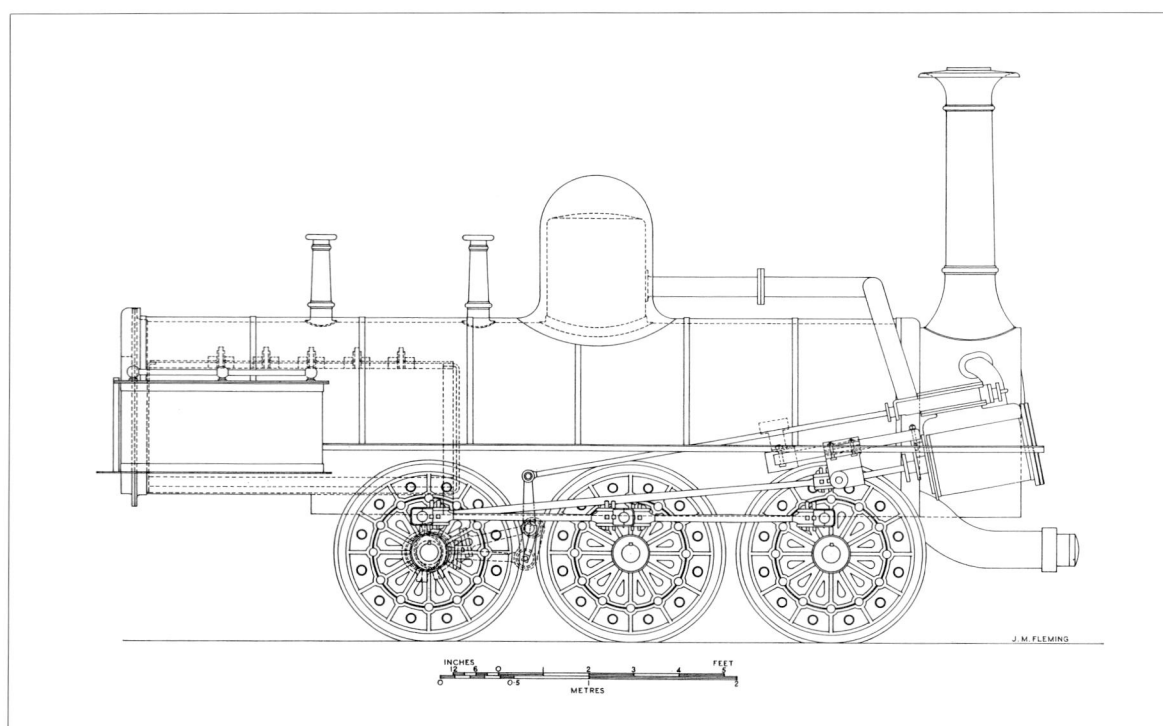

Fig 2.3
No 4 *Stockton* as remodelled in 1851.
(*J M Fleming*)

No 25, *Derwent*, built by A Kitching in November 1845, was one of four locomotives with 14½in by 24in cylinders supplied by that firm in 1845-8. Although similar in many respects to the Shildon-built *Miner* class they were never remodelled with fireboxes.

Five of the *Tory* class of locomotives of 1838-40 were remodelled with multi-tubular return-tube boilers and larger cylinders – two with 14½in by 20in cylinders in 1843 and three with 16in by 18in cylinders and link motion in place of the gab valve gear in 1853.

Locomotives of the *Dispatch* class were also fitted with larger cylinders, some being provided with link motion in addition. One of these was No 26, *Pilot*, built by W & A Kitching in 1840 and remodelled at Shildon Works in September 1852 (Fig 2.2). No 4, *Stockton*, was more extensively remodelled at Shildon in December 1851 with a firetube of Fossick & Hackworth's patent design. The overhanging weight of the lengthened boiler was counter-balanced by repositioning the cylinders on the smokebox (Fig 2.3).

No 35, *Commerce*, was one of three 0-6-0s for the merchandise traffic – Nos 35-37 – built to Bouch's designs at Shildon Works in 1847 (Plate 2.4). They had 16in by 24in outside cylinders placed horizontally with their transverse centres only 5ft 7½in apart; the connecting rods could thus be close to the driving wheels and the coupling rods on the outside – as on the locomotives with inclined cylinders – a large circular hole being formed in the connecting rods to clear the crank pins on the leading wheels. With these locomotives Bouch

Fig 2.4
No 38, *Rokeby*. 'Mecanicien', seen just in front of the leading wheels, was R H Inness' pen-name.
(*R H Inness*)

finally abandoned return-tube boilers and provided fireboxes designed to burn coke instead of coal.

S&DR 2-4-0 Passenger Locomotives 1847-1860

By 1847 there was need for additional locomotives for the S&DR passenger trains, the existing stock consisting of a miscellaneous collection of small locomotives acquired new or second-hand from a variety of sources.

The first additions were two 2-4-0s with bar frames - No 38, *Rokeby*, and No 39, *Ruby*, designed by Bouch and built at Shildon Works in June 1847 (Fig 2.4). They had 14½in by 24in outside cylinders spaced as close together as possible – as on the *Commerce* class 0-6-0s – to allow the connecting rods to be next to the wheels, the coupling rods being on the outside; the crossheads were made exceptionally wide and had vertical slots to clear the leading coupled wheel crank pins. The naves of the coupled wheels were made in two concentric parts, the annular gap between the castings being packed with hardwood and the whole secured with iron wedges; the small leading wheels had naves cast in one piece. All the wheels had round malleable iron spokes. Six-wheeled tenders with a capacity of 1,401 gallons of water and three tons of coke were provided.

Nos 38-39 were soon relegated to hauling merchandise trains. Both were provided with new cylinders in 1850, and the previously exposed wooden lagging on the boilers was covered with iron cleading. New wheels with flat spokes were fitted at an unknown date. No 38 was equipped with Bouch's feedwater heating apparatus in October 1857 (Plate 2.5).

Plate 2.4
Stockton & Darlington Railway. No 35, *Commerce*, built at Shildon in 1847 and photographed there at Mason's Arms crossing. *(Ken Hoole collection)*

The completion of Nos 38-39 coincided with the restoration of numbers to the earlier passenger locomotives, which were allocated new numbers between 40 and 55. Provision of brass numberplates was discontinued at the same time, all numbers being painted – probably in gold or yellow numerals – on the footplate side-sheets, domes and tenders. The practice of fitting brass nameplates on the sides of boilers continued for another twenty years.

Plate 2.5
Stockton & Darlington Railway. No 38, *Rokeby*, built at Shildon in 1847 but seen with a later cab, in place of its original weatherboard, and a jacketed chimney containing the feedwater heating apparatus installed in 1857. *(K L Taylor collection)*

Fig 2.5
No 58 *Woodlands*. (J M Fleming)

Reference to colours was made in a S&DR minute dated 1847, which mentions 'Brunswick green, Venetian red, Saxony green, scarlet lake and Indian red'. Unfortunately, no description was given of the actual arrangement of the colours. It is possible that Bouch introduced this new, more colourful livery for his latest passenger locomotives.

The passenger locomotives added to stock after 1847 included some useful 2-4-0s of more orthodox design, having a short coupled wheelbase and inside cylinders. The first two of these were No 58, *Woodlands*, and No 59, *Hall Garth*, built by W & A Kitching & Co in 1848 (Fig 2.5); their names were not allocated until 1852. One locomotive had a brass dome cover, the other having an iron cover with copper beadings; the handrails were brass. Seven more locomotives of generally similar dimensions, although not all alike in appearance, were built between 1855 and 1857, a final one being delivered in 1860.

No 71 was built by W & A Kitching to the maker's own designs and delivered to the S&DR in March 1851 (Fig 2.6); it was later named *Hackworth*. The coupled wheelbase was longer than that of the Woodlands class 2-4-0s, the cylinder castings extending right across the locomotive and being bolted directly to the outside frames. The valve chests were on the outer side of the cylinders, with the valve gear between the wheels and the outside frames. New cylinders with the valve chests between them were fitted in 1853, which allowed the leading wheels to be brought further forward.

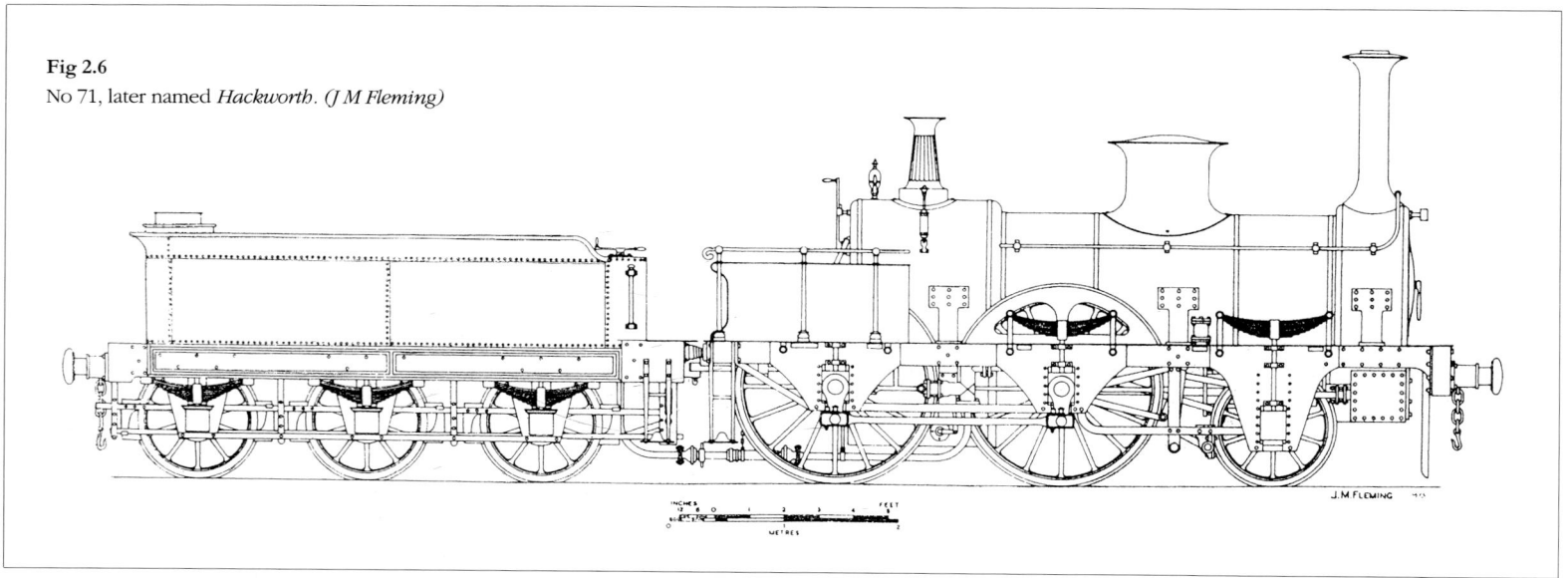

Fig 2.6
No 71, later named *Hackworth*. (J M Fleming)

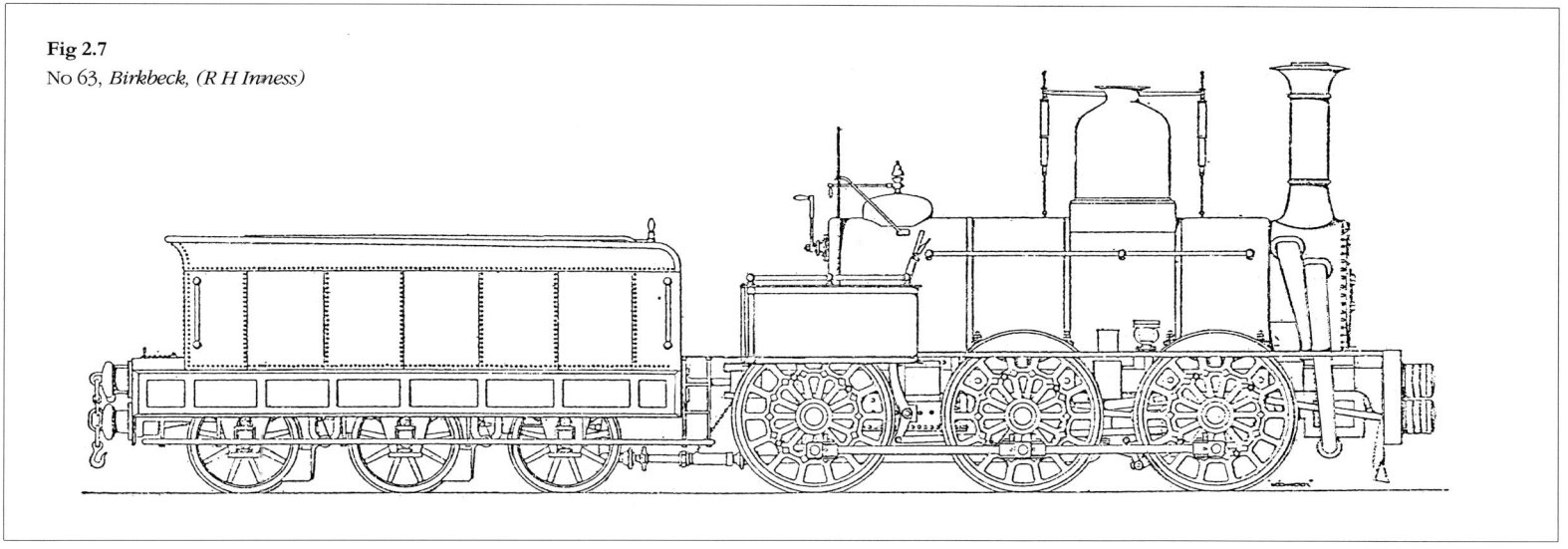

Fig 2.7
No 63, *Birkbeck*, (R H Inness)

S&DR Mineral and Merchandise Locomotives 1847-1853

Six 2-4-0s with outside cylinders, built for The Newmarket & Great Chesterford Railway which had got into financial difficulties, were taken over by the S&DR between 1847 and 1851, becoming Nos 65-70. They were assigned to the merchandise trains on which fast speeds were not required.

Seven 0-6-0s for mineral and merchandise traffic were built in 1848-9 to the designs of the S&DR Traffic Manager and Locomotive Running Superintendent, John Graham (Fig 2.7). They had main frames, inside cylinders and copper fireboxes, but several out-dated features from earlier designs were retained, including 'plug' wheels (more aptly described in many S&DR records as 'rim-and-centre' wheels) and external steam and exhaust pipes between the smokebox and the outside valve chests. Nos 63-64 were built at Shildon Works, Nos 60-62 by Gilkes, Wilson & Co and Nos 2-3 by Kitching & Co, the two outside firms each being charged £150 for the use of the Shildon drawings and patterns in the hope of achieving uniformity.

Nevertheless Nos 60-64 were built with 4ft 6in diameter wheels and 15in by 20in cylinders, whilst the final two – Nos 2-3 – had 4ft 0in diameter wheels and 15in by 20in cylinders. The locomotives ran without names until 1852, No 2 being then named *Graham*.

In their original form the boilers steamed badly and, in the late 1850s, those of Nos 60-64 were lengthened from 8ft $8^{1}/_{8}$in to 13ft. Nos 2-3 were withdrawn in 1865 without alteration.

The next six 0-6-0 mineral locomotives were of a much improved design incorporating inside plate frames. These were built by Gilkes, Wilson & Co in 1852-3, the first two being No 56, *Tow Law*, and No 57, *Shotley*, followed by Nos 72-75. They were not long-boiler (or short-coupled) locomotives, although described as such by previous writers, the boilers being 9ft 6¾in long by 3ft 10in diameter; the wheelbase of Nos 56 and 57 was 6ft 7½in + 6ft 0¼in, but ¼in shorter in the case of Nos 72-75. All had 4ft 2in diameter wheels and 17in by

Fig 2.8
No 72 *Peel*, (R H Inness)

Plate 2.6
Stockton & Darlington Railway. No 125, *Gazelle*, built by Gilkes Wilson in February 1858.

18in cylinders. Nos 72/3/5 were later remodelled with lengthened frames and boilers, the wheelbase being curtailed to 6ft 0in + 5ft 3in (Fig 2.8). Nos 56-7 were replaced by new long-boiler 0-6-0s in 1866.

S&DR Long-boiler Mineral Locomotives 1854-1863

Long-boiler 0-6-0s with inside plate frames and inside cylinders were introduced by William Bouch in 1854, the first two being No 76, *Prince of Wales,* and No 77, *Alexander*, both built by Gilkes, Wilson & Co in February of that year. They had 4ft 2in diameter wheels, with a single transverse spring for the trailing axle, and 17in by 18in cylinders, the wheelbase being 11ft 8in. The same firm built another 15 locomotives with similar dimensions between July 1854 and March 1856, Nos 83-85, 90-92, 95-97 and 102-107.

The long-boiler design proved to be very successful and remained the standard type of S&DR mineral locomotive until 1875, although with many variations in dimensions and appearance (Plate 2.6). Two early examples, built by R & W Hawthorn & Co and R Stephenson & Co respectively, are illustrated by Figs 2.9 & 2.10. The knuckle joint in the rear section of the coupling rods, giving some horizontal flexibility, was a feature of S&DR 0-6-0s for many years.

Nos 145-150, built by R & W Hawthorn & Co in 1860, were more powerful than the earlier long-boiler 0-6-0s, their wheels being 5ft 0in diameter and the cylinders 17in by 26in (Fig 2.11). They were ordered in preparation for the opening of the new line from Barnard Castle to Tebay, which included severe gradients on the long climb to Stainmore Summit, 1,370ft above sea level. Large side-window cabs were originally provided to shelter the enginemen from the exposed conditions on the high moorland, but these were later replaced by Bouch's usual style of cabs. The locomotives had Bouch's feedwater heating apparatus, in which the right hand feed pump delivered cold water into the top of a heater jacket surrounding the chimney, the heated water being drawn off at the base and pumped via the left-hand clackbox into the boiler. The apparatus could be by-passed to allow both feed pumps to deliver cold water if desired.

Six more locomotives for the Stainmore line – Nos 151-156 – were built by Gilkes, Wilson & Co between June 1860 and February 1861, these having 5ft 0in diameter wheels and 17in by 24in cylinders. Another seven were built between August 1861 and January 1864 – Nos 157-159 and 167-170 – and two more – Nos 187-188 – in 1864-5. (No 157, *Planet*, was built with 4ft 2in diameter wheels). Incidentally, No 169, *Tufton*, built in May 1863, was the last locomotive delivered to the S&DR before the amalgamation with the NER. No 170, *Reliance*, is illustrated by Plate 2.7.

S&DR Bogie Passenger Locomotives 1860-1862

In 1860 R Stephenson & Co. built two 4-4-0s with outside cylinders for the passenger traffic over the Stainmore line, No 161, *Lowther*, being delivered in August and No 160, *Brougham*, in October (Plate 2.8 and Figs 2.12 & 2.13). The low-pitched boilers of these locomo-

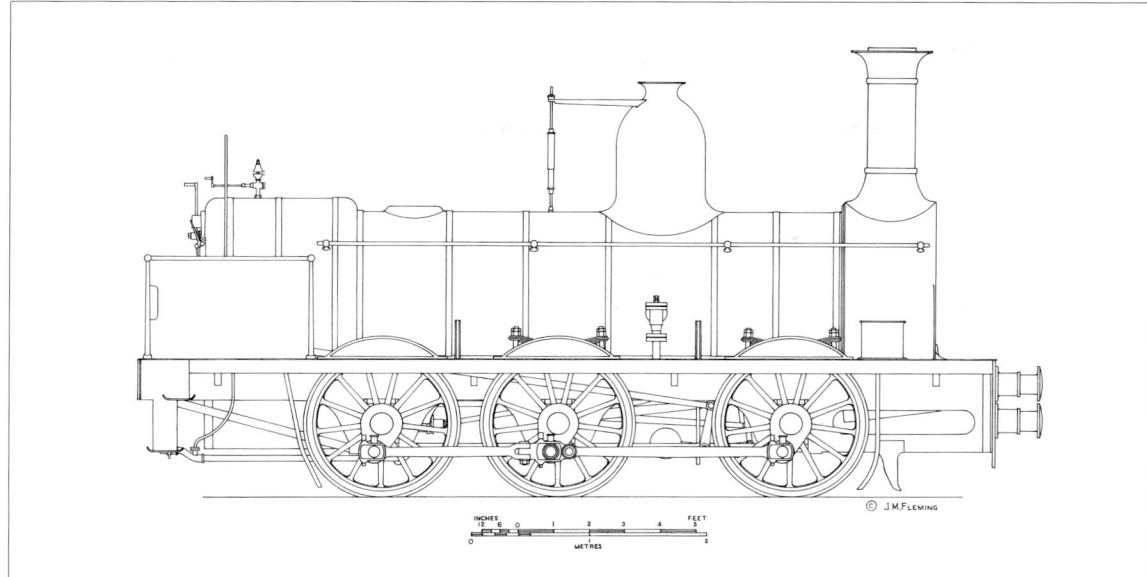

Fig 2.9
No 129, *Stanley,* built by R&W Hawthorn in April 1858. *(J M Fleming)*

Fig 2.10
No 136, *Tebay,* built by Robert Stephenson & Co in November 1858. *(J M Fleming)*

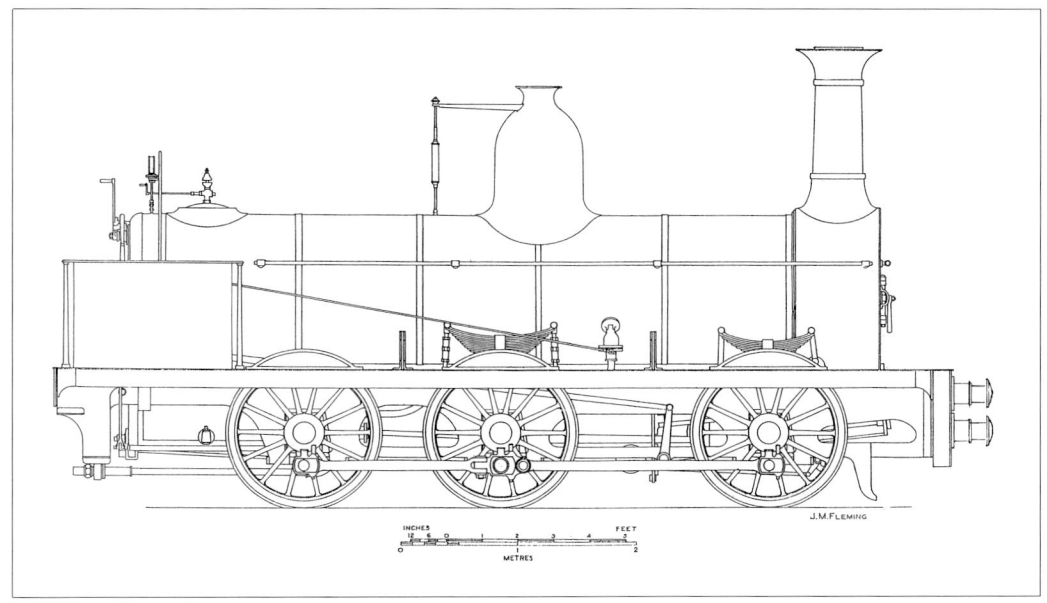

Fig 2.11
No 145, *Panther,* supplied by R & W Hawthorn in January 1860 and seen with its original cab. *(R H Inness)*

tives gave them a very sleek and elegant appearance, which was enhanced by their distinctive cabs.

No contemporary description of their original livery has been found but the main body colour would probably have been Saxony green, picked out with black bands and fine white lines. The cylinders, footplate angle-irons and tender frames were a darker colour – probably chocolate.

Four more 4-4-0 passenger locomotives – Nos 162-165 – were built to a modified design by R Stephenson & Co in 1862, the diameter of the coupled wheels being increased to 7ft 0½in (Plate 2.9 and Fig 2.14). The bogies could turn about a fixed pivot but there was no provision for lateral movement. These locomotives were fitted with

Plate 2.7
Stockton & Darlington Railway. No 170, *Reliance*, completed by Gilkes Wilson in January 1864. *(Ken Hoole collection)*

Bouch's steam brake, the equipment consisting of a steam cylinder fixed between the coupled wheels which operated a pair of hinged iron bands fitted with wooden brake blocks; application of steam caused these blocks to be pressed down on the top of the wheels and apply a braking force.

Only a low weatherboard was provided for the protection of the men, the large cabs of the previous 4-4-0s having proved unpopular. In addition to the ornamental brasswork on the wheel splashers there were narrow brass beadings around the footplate side-sheets and weatherboard.

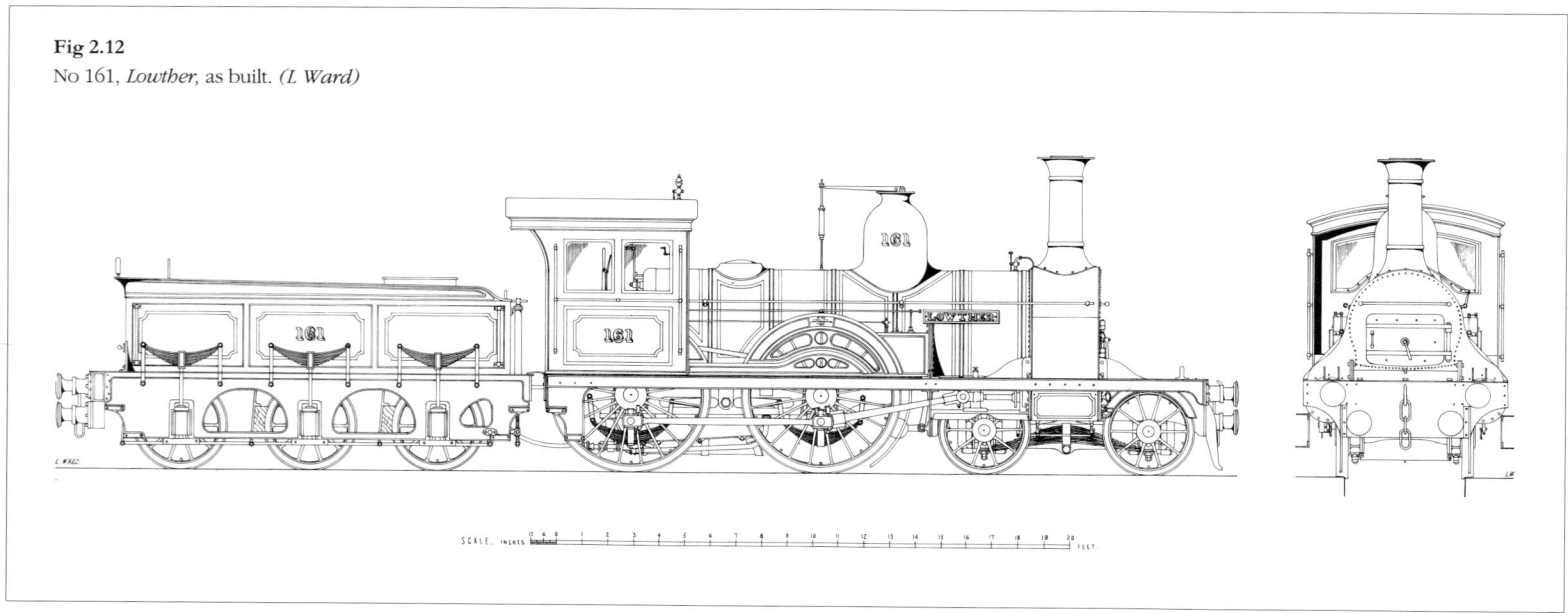

Fig 2.12
No 161, *Lowther*, as built. *(L Ward)*

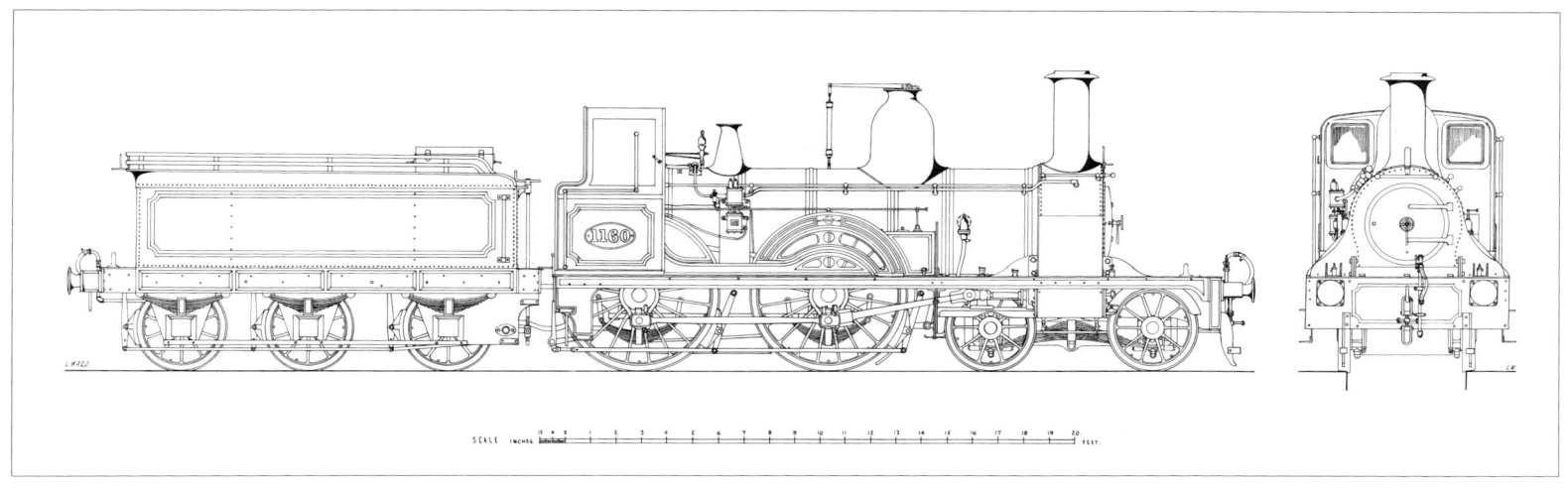

Fig 2.13
No 160, *Brougham* as rebuilt and shown carrying its NER number 1160. *(L Ward)*

Fig 2.14
No 163, *Morecambe. (J M Fleming)*

Plate 2.8
Stockton & Darlington Railway. No 160, *Brougham*, seen in front of Alpha Place, Saltburn, a common setting for posed photographs of S&D locomotives. The capacious cab was soon replaced by the normal S&D variety. *(Ken Hoole collection)*

Plate 2.9
Stockton & Darlington Railway. No 165, *Keswick*, at Saltburn.

The Combined Locomotive Stock of the NER and S&DR

At the time of the amalgamation with the NER in June 1863 the S&DR had 157 locomotives, three of these being classed as "occasionally available" and three as "old". The NER with 491 locomotives had slightly more than three times as many as the S&DR. The former S&DR locomotives and the subsequent additions built for the Darlington Division were not renumbered in the NER list until October 1872.

Plate 2.10
West Hartlepool Harbour & Railway. A typical later 0-6-0, No 47, built by them in 1863 with 4ft 10in diameter wheels and seen as NER No 617 with the distinctive Hartlepool cab.

The West Hartlepool Harbour and Railway (WHH&R)

The incorporation of the N&CR and S&DR within the NER was followed soon afterwards by absorption of the WHH&R on 30 June 1865, the latter having been formed in 1853 by an amalgamation of two early railways, the Clarence and the Stockton & Hartlepool.

The WHH&R favoured 0-6-0s with outside cylinders for its mineral traffic, many of these being of a curiously out-dated design with cylinders mounted obliquely on the sides of the smokebox and connecting rods driving the rear pair of wheels. After 1862 a change was made to horizontal cylinders (Plate 2.10). It is known that many locomotives had names as well as numbers, some of the nameplates surviving for several years after the locomotives had been renumbered by the NER. The locomotive livery at one time was light green picked out with fine yellow lines.

The WHH&R had no influence on NER locomotive design if one excludes the 'Hartlepool' style of angular-shaped cabs added to some locomotives, e g 0-6-0 No 617 and Fletcher's 0-6-0 No 160 (Plate 1.16). The former WHH&R Works continued to repair locomotives for 14 years after the amalgamation.

At the date of the amalgamation the WHH&R owned 69 locomotives with four more under construction at the company's workshops. By November 1866 only 61 remained in stock, 58 of which were allocated NER numbers between 584 and 641, whilst three became Nos 19-21 in the stock of the Darlington Division.

Plate 2.11
West Hartlepool Harbour & Railway. The company had a number of short-wheelbase, four-coupled tank locomotives for working tight curves at the docks. 0-4-2T No 72, *Samson*, was built by Fossick & Hackworth in 1861 with 4ft diameter coupled wheels. Seen here as NER No 640, it was scrapped in 1881.

Plate 2.12
West Hartlepool Harbour & Railway. No 41, *Alexandra*, seen here as NER No 613, was built in December 1865 with 3ft 6in diameter wheels. It was renumbered 1777 and 1788 before being scrapped in February 1894. It is seen here in front of the staiths at West Hartlepool.

Chapter 3

NER Locomotive Progress under Two Locomotive Superintendents 1863 – 1869

From 1863 to 1876 the record of locomotive progress is complicated by the fact that developments on the Darlington Division, under William Bouch, proceeded independently of contemporaneous developments at Gateshead Works under Edward Fletcher. To avoid confusion as far as possible, the successive new locomotive designs and liveries introduced by each of the two Locomotive Superintendents during the 'independent' period are described in separate narratives, both of which are sub-divided into two periods, namely 1863-1869 and 1870-1876, although not without an occasional overlap.

Passenger Locomotives 1863 - 1869:
Edward Fletcher, Locomotive Superintendent

Three small 2-4-0s were built at Gateshead Works in 1863, two of which were tried on the Whitby branch. Having a very short wheelbase for negotiating the numerous curves on the line, they were found to be too unsteady for working the passenger trains and were soon transferred elsewhere.

More successful were the ten 4-4-0 'Whitby Bogies' – Nos 492-501 – built by R Stephenson & Co in 1864-65 (Fig 3.1). These were Fletcher's only tender locomotives with leading bogies and were intended for working throughout from Whitby to Malton after the new 'deviation line' had replaced the rope-worked Goathland incline in July 1865. A few months later a new section of line between Grosmont and Castleton allowed through running from Whitby to Stockton.

The 4-4-0s had substantial outside main frames formed from iron plates $1^{3}/_{8}$in thick and additional inside frames and bearings for the crank axles. The bogies originally had a short laminated spring below each axlebox, the two on each side being linked by equalising levers, but these were soon superseded by an unusual arrangement of outside springs and axleboxes in conjunction with the inside frames (Plate 3.1 and Fig 3.2). The 'Whitby Bogies' were the first of Fletcher's locomotives to have tapered 'stovepipe' chimneys with only a narrow beading around the top instead of an ornamental cast-iron cap. (The 1886 designation was Class 492.)

While Fletcher's 2-4-0s with outside frames and bearings (Plate 3.2) were still being built at Gateshead Works, an order for ten 2-4-0s of a different design was being delivered from R & W Hawthorn & Co. The latter comprised Nos 544-553 (Plate 3.3) built between September 1865 and January 1866, which had 'mixed' frames, i.e. outside for the leading axle and inside for the coupled pair, the wheelbase being

Fig 3.1
Whitby Bogie No 500, as built. *(L Ward)*

Plate 3.1
Class 492. Whitby Bogie No 500, built in April 1865 and seen at Malton in its original condition except that a single long equalising spring on the front bogie has replaced individual springs on each axle. It was scrapped in February 1885, a new No 500 having been completed the previous September.

Plate 3.2
Class 25. The later appearance of No 26, built at Gateshead in January 1865, seen after receiving a Fletcher cab in place of its original weatherboard and being reboilered in 1883.
(Ken Hoole collection)

Plate 3.3
Class 544. No 545 was built by R & W Hawthorn (Works No 1308) in October 1865 and is seen in its original condition. It was rebuilt in 1881, renumbered 1939 in January 1892 and 1753 in January 1894 before withdrawal in August 1904.

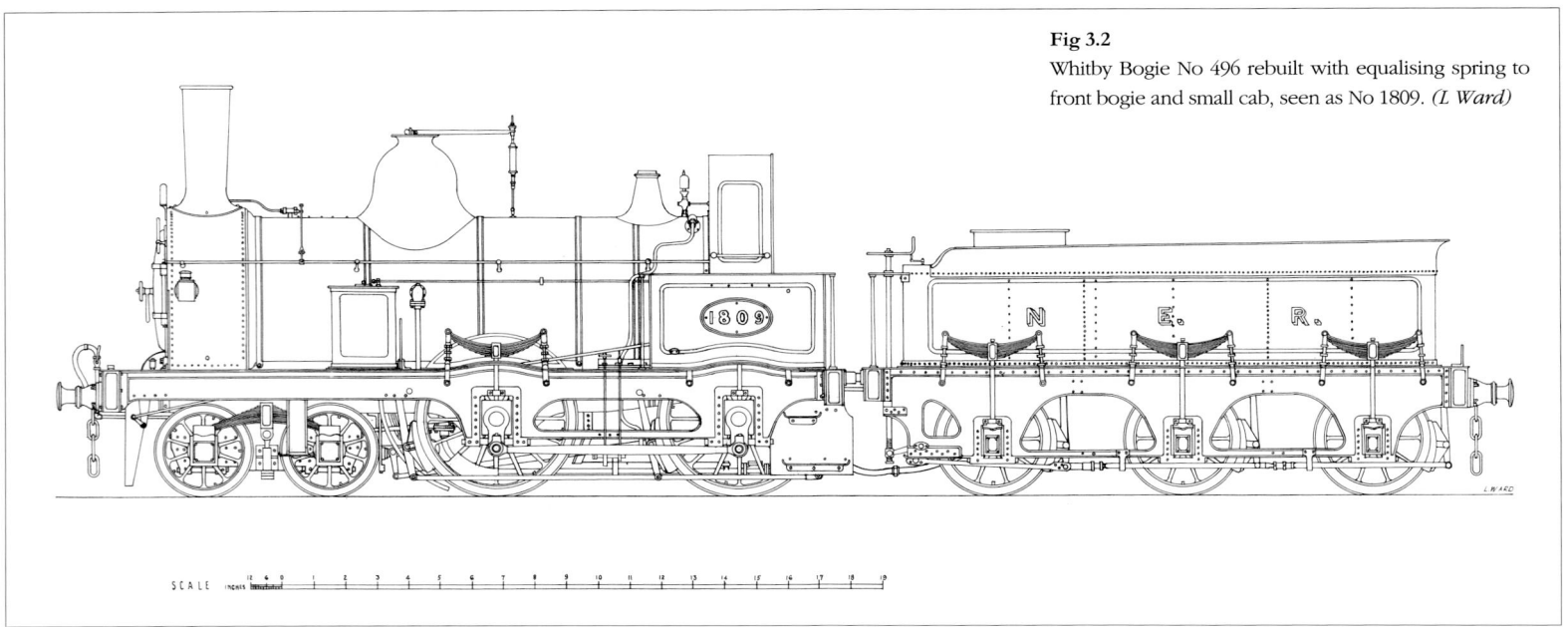

Fig 3.2
Whitby Bogie No 496 rebuilt with equalising spring to front bogie and small cab, seen as No 1809. *(L Ward)*

the same as that of the Gateshead 2-4-0s. The leading dimensions of the two classes were similar and they shared most of the main-line passenger workings until displaced by larger locomotives after 1872. The Hawthorn locomotives worked mainly from York whereas the Gateshead locomotives were used north of Newcastle. (The 1886 designation was Class 544.)

A smaller version of the outside-framed 2-4-0s was brought out in 1867, the first one being No 38. The same design of outside sandwich frames was used in conjunction with 6ft 0in diameter coupled wheels and a 4ft 0in diameter leading pair, the height of the running plates being lowered by three inches; the wheelbase was unchanged, namely 7ft 10½in + 8ft 0in. Eight of these locomotives were built between July 1867 and October 1870. (Note: a small 2-4-0 with 6ft diameter wheels, bearing a spurious '38' numberplate, is illustrated on page 45 of J.S.MacLean's *The Locomotives of the North Eastern Railway: 1841-1922*. This was a copy of an old Gateshead Works drawing which is now known to have been an earlier proposed design dating from the 1850s)

Similar outside sandwich frames were also used for three 2-4-0s with 5ft 6in and 3ft 6in diameter wheels that were built in 1869 for the steeply-graded branch from Blaydon to Blackhill. The Gateshead Works drawing shows that they were to have had bent-over weatherboards with added side panels to form rather primitive cabs, but at least one of the trio was built with an early version of Fletcher's round-topped cab. (These two varieties of 2-4-0 were later included in Class 25.)

A small 2-4-0, No 152, rebuilt at Gateshead in May 1868 (Fig 3.3), is of interest because of a drawing and brief note of its livery provided by W G Brown, who had driven the locomotive when new. He described the basic colour of the locomotive and tender as green, including their outside frames, the inside frames and cylinders being brown: the number plates and buffer beam were red and the smokebox and chimney were black. Unfortunately, he omitted to give a precise description of the shades.

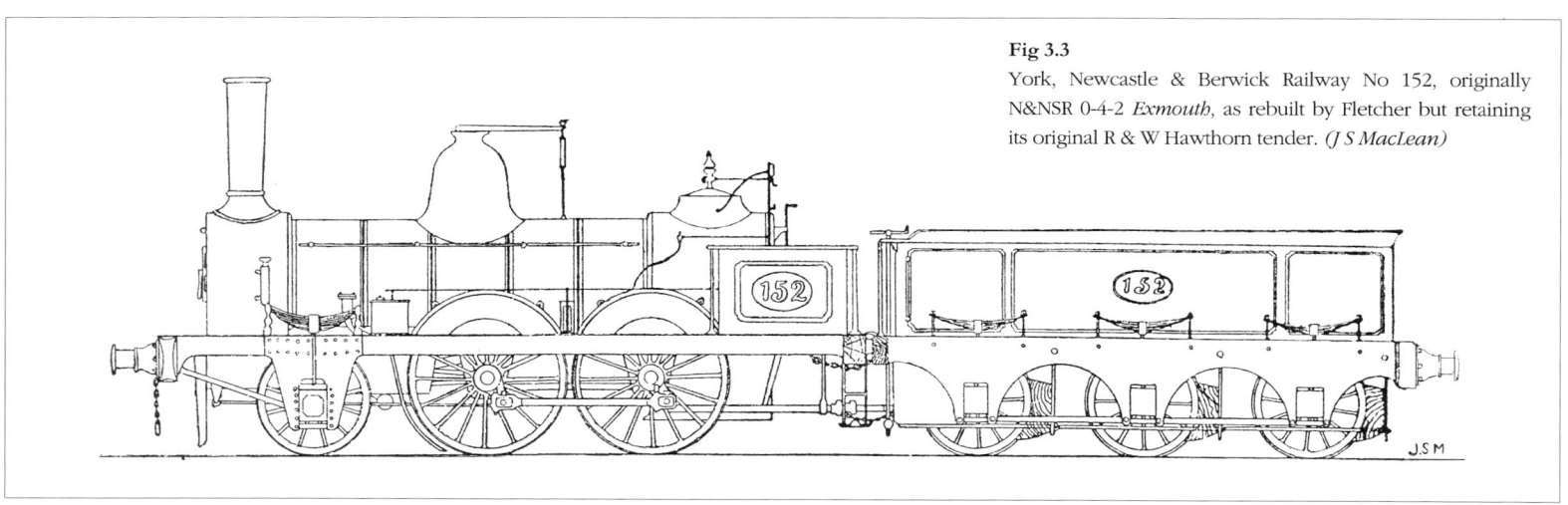

Fig 3.3
York, Newcastle & Berwick Railway No 152, originally N&NSR 0-4-2 *Exmouth*, as rebuilt by Fletcher but retaining its original R & W Hawthorn tender. *(J S MacLean)*

Plate 3.4
Class 13, No 531, built by R & W Hawthorn. A flat plate securing the boiler to the frames can be seen between the leading and middle splashers. *(Ken Hoole collection)*

Plate 3.5
Class 93. No 658, built by Robert Stephenson & Co. The date of the photograph has been marked on the tender, by scribing the photographic plate.

Goods Locomotives 1863 - 1869:
Edward Fletcher, Locomotive Superintendent

The goods locomotives added to stock during the period 1863-69 included Nos 524-543, built by R & W Hawthorn & Co between 1865 and April 1866, and Nos 554-558 by Hudswell, Clarke & Co between August 1865 and April 1866, both lots having iron sandwich-pattern outside frames. The original appearance of No 531, built in August 1865, is illustrated by Plate 3.4.

R & W Hawthorn & Co also supplied short-coupled 0-6-0s Nos 502-509 between August and November 1864, followed by Nos 564-583 between June and November 1866. These had wooden sandwich-pattern outside frames, the boilers of the earlier batch being attached to the frames by flat brackets – as seen on No 531 – while the boilers of the later locomotives were supported on transverse iron plates.

The contemporary short-coupled 0-6-0s built by R Stephenson & Co had outside frames formed from 1in thick iron plates. Nos 510-517 were built between December 1864 and August 1865, followed by Nos 642-661 between November 1866 and January 1868. No 658 is illustrated as it appeared in 1881, after the provision of brakes (wooden brake blocks), a safety valve over the firebox and with sides added to the weatherboard (Plate 3.5)

The short-coupled mineral locomotives with inside frames, added during the 1860s, were all built at Gateshead. A drawing of the frames for these 0-6-0s, dated October 1869, shows a longitudinal iron beam, 10in by 1in, with hornplates riveted on each side, the wheelbase being 5ft 11in + 5ft 3½in.

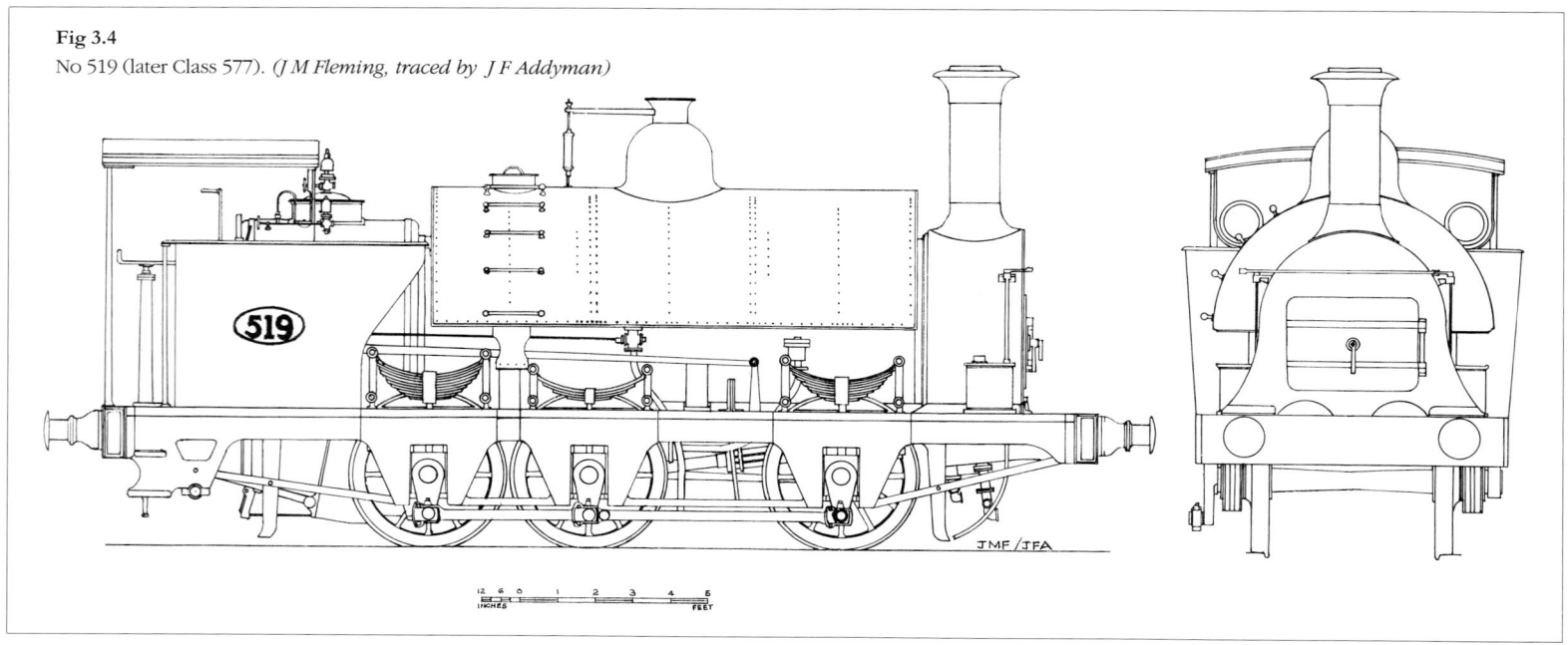

Fig 3.4
No 519 (later Class 577). *(J M Fleming, traced by J F Addyman)*

Tank Locomotives 1863 - 1869:
Edward Fletcher, Locomotive Superintendent

The NER had very few tank locomotives before the 1860s when several were obtained for certain specific duties. In 1864 two short-coupled 0-6-0 saddle tanks with outside frames – Nos 518-519 – were built by Manning, Wardle & Co for assisting trains up inclines on the Whitby branch (Plate 3.6 and Fig 3.4). They had 4ft 0in diameter wheels and 16½in by 24in cylinders, the wheelbase being 6ft 8½in + 4ft 3½in. At the same time the firm had under construction for the NER two long-boiler 0-6-0 tender locomotives with 5ft 0in diameter wheels, Nos 520-521, which were similar to No 93 built at Gateshead four years earlier. Although Nos 518-19 had the same design of boiler but smaller wheels and a shorter wheelbase, they could be said to be the saddle-tank version of the 0-6-0s. Some years later, four of these 0-6-0s, Nos 572, 575, 576 and 577, which had been built by other contractors, were converted to 0-6-0 saddle-tanks and they, along with Nos 518-19, became Class 577 under the 1886 designation.

Three large 0-4-0 saddle tanks – Nos 662-664 – were built by Hudswell, Clarke & Co in 1866-67 for use at Tyne Dock; they had

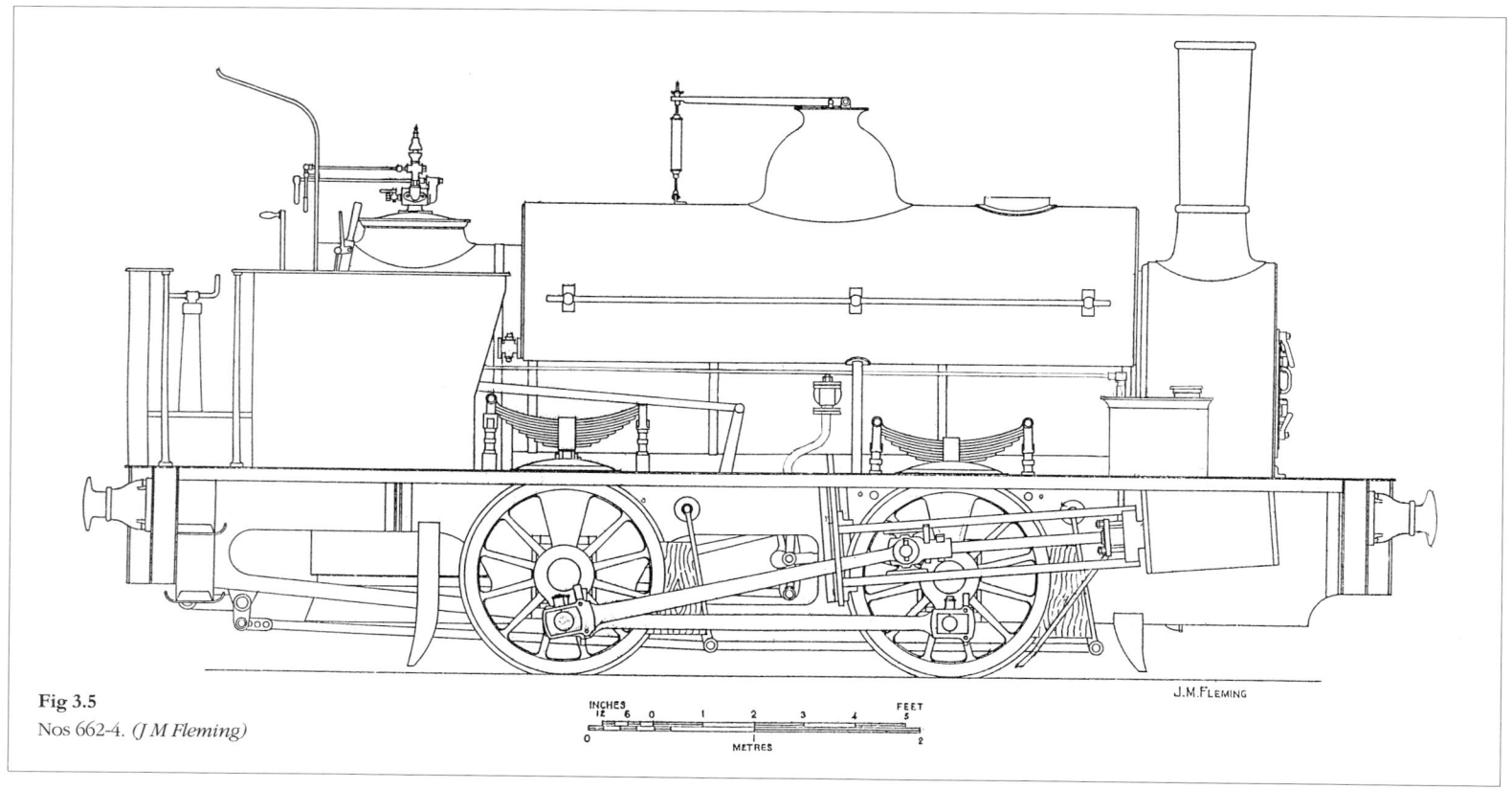

Fig 3.5
Nos 662-4. *(J M Fleming)*

Fig 3.6
No 20, with the York, Newcastle & Berwick Railway numberplate from the locomotive which it replaced. *(J M Fleming).*

4ft 0in diameter wheels and 16 in by 24in outside cylinders (Fig 3.5). Between 1871 and 1873 they were all rebuilt as 0-4-2 side tanks, their discarded tanks being used later to convert three long-boiler 0-6-0s from tender to tank locomotives.

Gateshead Works built two short-coupled 0-6-0 saddle tanks in 1869 which, although new locomotives, were given the old YN&BR numberplates – Nos 5 and 20 – off the locomotives that they replaced (Fig 3.6 and Plate 3.7). They had 4ft 2in diameter wheels, 14½in by 22in cylinders and 6ft 6in + 5ft 2¼in wheelbase.

From 1866 the locomotives built at Gateshead carried 'works plates' that showed the company's name, the name of the works and the building date. The early works plates, used between 1866 and 1872, were brass castings with raised letters and figures, and were almost always painted - not polished (Plate 3.8).

Plate 3.6
Class 577. No 519 built by Manning, Wardle in 1864. *(Ken Hoole collection)*

Plate 3.7 (above)
No 5 is one of a pair of 0-6-0ST (the other being No 20) built at Gateshead in October 1869. The numberplate has been re-used from the previous No 5; below it is the works plate. A primitive cab has been created by adding side plates to the original weatherboard.

Plate 3.8 (right)
Gateshead works plate of period 1866-72. Detail from Plate 3.7

Darlington Division Mineral Locomotives 1863 - 1869: William Bouch, Locomotive Superintendent

The new North Road Engine Works of the S&DR, at Darlington, were opened on 1 January 1863 but dealt only with repairs and rebuilding until 1864. Meanwhile, after an interval of five years, construction of new locomotives had recommenced at Shildon Works, beginning with a series of long-boiler 0-6-0s having 4ft 11in diameter wheels, 17in by 24in cylinders and a wheelbase of only 6ft 0in + 5ft 3in. Nos 171-2 were completed in 1863, Nos 173-4 in 1864 and Nos 189-90 in 1865.

The first entirely new locomotive built at North Road Works was No 175, *Contractor*, completed in October 1864 (Plate 3.9). It was similar in many respects to the foregoing Shildon-built locomotives but had a 6ft 5in + 5ft 3in wheelbase and an improved form of Bouch's valve gear. The distinctive cast-iron caps on the domes, seen in the illustrations of Nos 175 and 182, were fitted to most Darlington mineral locomotives until 1870; plain 'stovepipe' chimneys had begun to supersede those with bell-mouthed caps a few years earlier.

No 175 had a ¾in wide brass beading along the lower edge of the running plate angle-irons in addition to the beadings around the front and back edges of the cab. The rectangular border painted around the locomotive's number was a short-lived addition to the usual Darlington livery around 1864.

Ten 0-6-0s of a new and enlarged design – Nos 176-185 – were built by R Stephenson & Co between April 1865 and April 1866 (Plate 3.10). The wheels were 5ft 0½in in diameter, the cylinders 17in by 26in and the wheelbase 7ft 9in + 5ft 3in; the boiler barrel was 14ft in length, the raised firebox casing being 5ft 3in long and having an outwardly flanged backplate. Bouch's feedwater heating apparatus was provided and the locomotives also had Bouch's steam brake operating on top of the driving and trailing wheels. Their tenders had 4ft 0in diameter wheels, an 11ft 0in wheelbase, and a capacity of 1,800 gallons of water and 3½ tons of coal. Two locomotives of similar design were built at North Road Works, these being No 186, *Union*, in February 1865 and No 191, *Autumn*, in April 1866.

Plate 3.9
Darlington Division. No 175, *Contractor*, posed outside Darlington North Road Works, where it was the first locomotive to be completed.

Plate 3.10
Darlington Division. No 182, *Elton*, built by Robert Stephenson & Co in December 1865, equipped with Bouch's feedwater heating system, seen outside Shildon roundhouse.

North Road Works reverted to 4ft 2½in diameter wheels for eight 0-6-0s built between December 1865 and October 1866. Nos 2, 3, 56 and 57 replaced old locomotives but did not inherit their names: Nos 192-195 all bore names, the last two being erected at Shildon Works from parts mainly manufactured at North Road. Four somewhat similar 0-6-0s with 5ft 0in diameter wheels were built at North Road between March and May 1867; Nos 27 and 53 were only given numbers but No 93, *Uranus*, (Plate 3.11) and No 94, *Neptune*, also took the names of the locomotives that they replaced. No 202, *Ireland*, and No 203, *England*, built at Shildon in June 1867, were the last new locomotives turned out from the old works and also the last Darlington Division locomotives to receive names.

At the end of 1866 four short-coupled 0-6-0 saddle tanks – Nos 196-199 – were supplied by R Stephenson & Co for working on the Skinningrove zigzag incline, although at least two worked initially on the Stainmore line; both were involved in accidents in 1867 (No 199 on 5 January and No 196 on 6 April). Lacking the constraint provided by tenders these long locomotives were found to ride badly and damage the track: in October 1867 there was a proposal to add an extra pair of wheels but that scheme was not adopted. Additional water tanks were fitted at each side of the smokebox of No 199 (Fig.3.7) but evidently failed to steady the locomotive and eventually Nos 196-199 were all converted to tender locomotives in 1871-72. (The rebuilds were later designated Class 1196.)

Plate 3.11
Darlington Division. No 93, *Uranus*, standing outside North Road Works, where it was built in 1867. *(Ken Hoole collection)*

Between November 1865 and April 1868 Hopkins, Gilkes & Co built fifteen 0-6-0s with 5ft 0in diameter wheels, ten of these replacing old locomotives and five – Nos 200/1/4/5/6 – being additions to stock. Five more – Nos 226-230 – followed between September 1870 and April 1871 (Fig 3.8), their boilers having dome covers of the already obsolescent pattern with cast-iron caps. An early photograph of No 229 in service – not reproduced – shows the locomotive painted dark slate grey, with numerals on the cab side and dome in white.

Twelve large 0-6-0s fitted with Bouch's feedwater heating apparatus and steam brake were delivered by R & W Hawthorn & Co between December 1867 and May 1868, their numbers being 207-218 (Plate 3.12). The dimensions of the wheels, cylinders and wheelbase were the same as those of the large 0-6-0s Nos 176-185 but the boilers were 14ft 6in long and had flat-topped firebox casings, the latter being 5ft 3in long by 5ft 0in wide at the top and 4ft 0in wide below. Large tenders with a capacity of 1,800 gallons of water and three tons of coal were provided. Nos 207-218 were to have been given the names of the twelve months of the year, but they remained nameless.

Nos 221-225, built to the same design except for having 17in by 28in cylinders, were delivered by Hawthorns between August and November 1870. The names of Nos 221-224 were to have been the days of the week from Wednesday to Saturday respectively, Sunday

Fig 3.7
No 199, designed for working the Skinningrove Incline, showing the extra water tanks flanking the smokebox. *(R H Inness)*

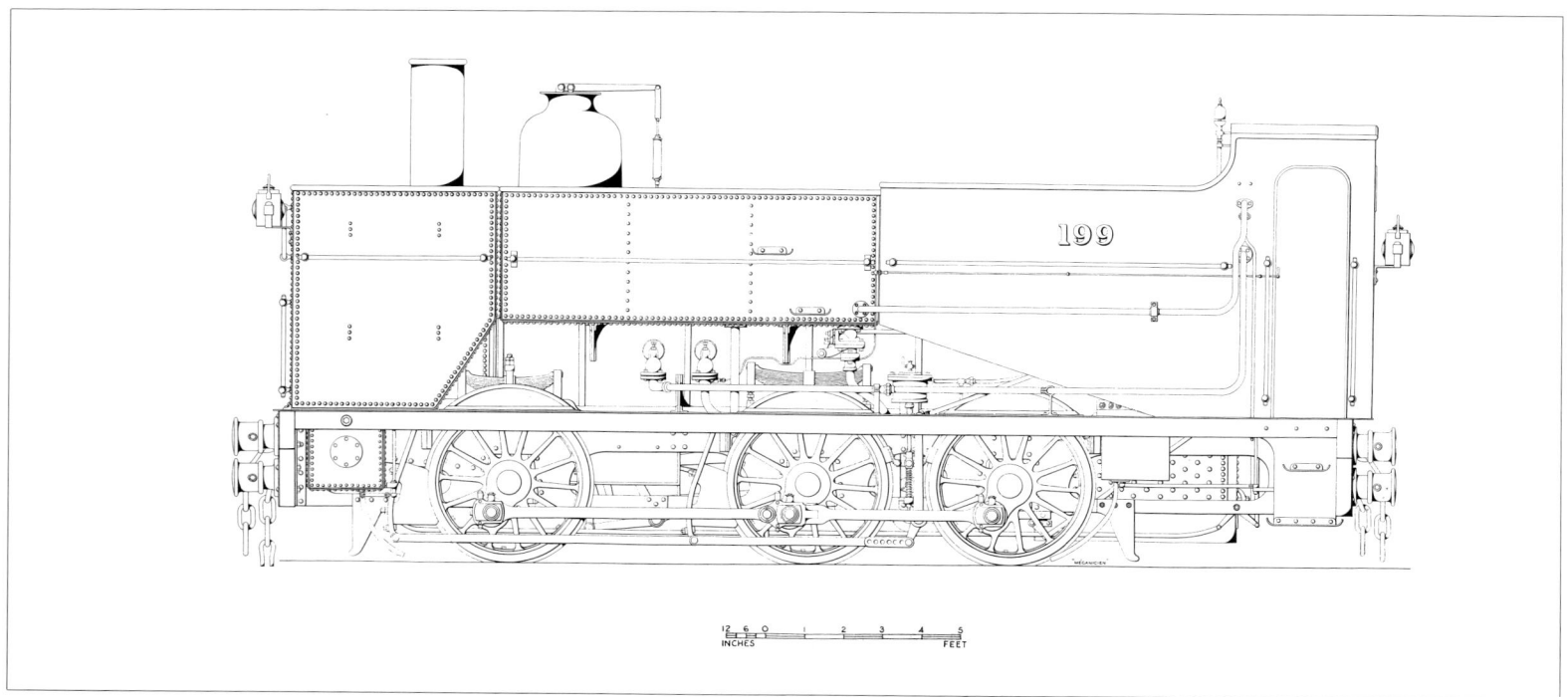

52

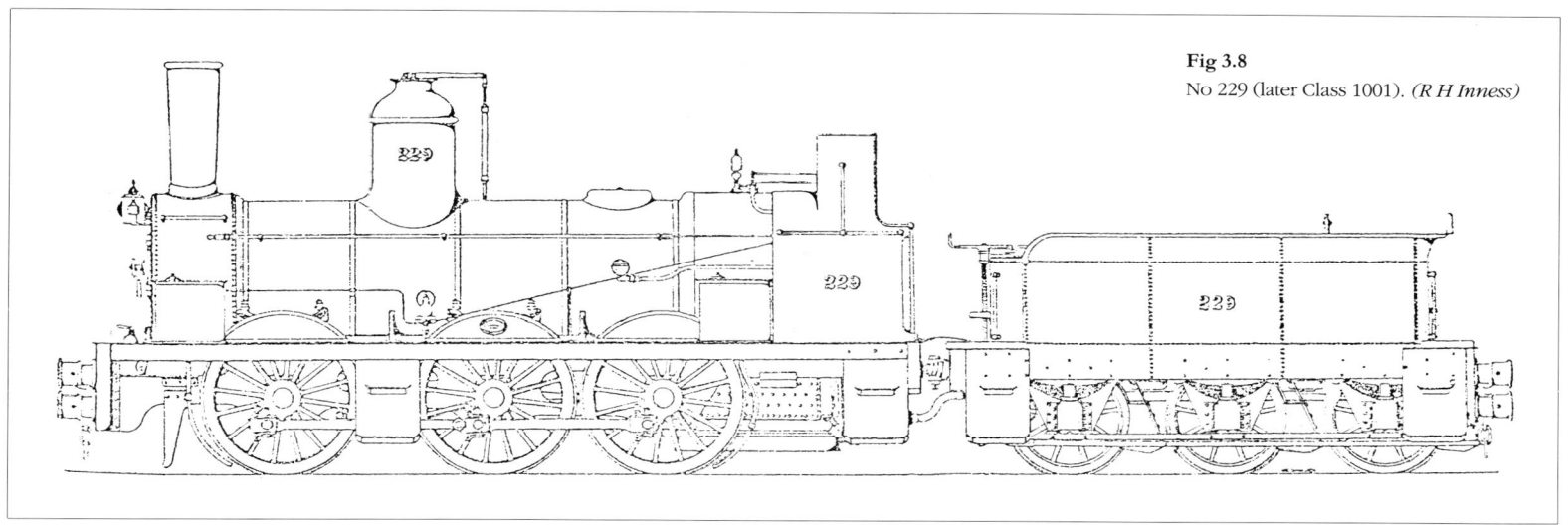

Fig 3.8
No 229 (later Class 1001). *(R H Inness)*

was evidently considered to be an inappropriate name for a locomotive. In the event, they remained unnamed.

Nos 1207/8/10/11/23 were rebuilt between 1877 and 1879 with a longer wheelbase and shortened boilers with round-topped fireboxes (their original numbers had been increased by 1000 in 1872).

The first 29 examples of what became the Darlington Division's 'standard' class of long-boiler mineral locomotive – later designated Class 1001 by the NER – were built at North Road Works between April 1868 and December 1870 (Plate 3.13). They had 5ft 0½in diameter wheels and a 6ft 7in + 5ft 3in wheelbase, the majority having 17in by 24in cylinders, although 17in by 26in had been adopted by June 1870. The boilers were 14ft long by 4ft diameter internally, the firebox casings being 4ft 8¾in long; tall, bell-shaped dome covers superseded the older pattern in 1870. None of these locomotives actually carried a name, although *Monday* and *Tuesday* had been selected for Nos 219 and 220 respectively; all the others were given the numbers of the old locomotives which they replaced. Among the superseded engines was No 1, *Locomotion*, of 1825, which was at last replaced by a new No 1 built in May 1868. The new No 1 was renumbered 1001 by the NER after October 1872, was reboilered in 1894 and again in 1904, but did not survive a wholesale withdrawal of old locomotives in November 1905.

Plate 3.12
Darlington Division. No 214, supplied by R & W Hawthorn in March 1868.
(J F Mallon collection)

Plate 3.13
Darlington Division. Class 1001 No 7, built at Darlington in June 1869.

Chapter 4

NER Locomotive Progress under Two Locomotive Superintendents 1870 – 1876

Edward Fletcher, Locomotive Superintendent

The period from 1870 to 1876 was notable for the introduction of a large number of locomotives built to new and improved designs. Edward Fletcher finally abandoned built-up inside frames in 1870 in favour of iron plate frames, and no more locomotives with outside sandwich frames were built after 1873. The output of new locomotives from Gateshead Works was augmented by no less than 392 built to Fletcher's specifications by outside manufacturers and, in addition, 15 tank locomotives were supplied by four different firms to their own designs.

The improvements in design and construction were accompanied by a change to a much darker green livery which was used by Gateshead paintshop until *c*1878. W B Thompson, who was familiar with the locomotives seen at Carlisle in the 1870s, described a few of them thus:- "I do not remember 185 (2-4-0 built in 1848) showing a trace of paint at all. Certainly she was very dark but whether painted black or merely very dirty I cannot say. Other locomotives, e.g. 440 (0-6-0 built in 1861) and 396 (0-6-0 built in 1854) were in the same condition. About 1873 we began to get the Stephenson goods locomotives of the 398 class (0-6-0s 879-897 built in 1873) down at Carlisle. As far as I remember they were originally a very dark green with the inside of the cab a bright red. Then the Sharp, Stewart 1395 class (0-6-0s 1390-1409 built in 1875) were black or practically black like the LNWR." [1]

The above description is confirmed by early photographs of locomotives built in 1870-78. The elaborate panelling and lining-out seen on the principal passenger locomotives and some of the newer goods locomotives was not universal; less favoured goods locomotives were dark green with only narrow black borders and black panel bands.

A hint of the probable livery of locomotives painted at York Works is given by the wooden model of a *Jenny Lind* 2-2-2 which was made by workmen at the carriage works in 1871 to commemorate the granting of a nine-hour working day. The model, now in the National Railway Museum, is a fairly accurate representation and is painted thus:- Boiler, footplate side-sheets and tender body – darkish green with black cleading belts and panel bands, edged white: Outside frames, ends of buffer beam, splashers and sandboxes – chocolate, with black edges lined white: face of buffer beam – vermilion with black borders and white lining. The brass numberplates were vermilion with polished numerals and borders.

Plate 4.1
Class 686. No 696, supplied by Beyer Peacock in 1870 and bearing the maker's name around the splasher.

Passenger Locomotives 1870 - 1876:
Edward Fletcher, Locomotive Superintendent

Twelve 2-4-0s for main-line traffic – Nos 686-697 – were built by Beyer, Peacock & Co in 1870 (Plate 4.1). These had 6ft 0in diameter coupled wheels and 16in by 22in cylinders, substantial plate frames, a boiler with raised firebox and a Naylor safety valve near the front of the barrel in addition to the usual pair of spring balance valves on the dome; they also had a new style of cab. The tenders had wooden sandwich frames with iron flitch plates and separate horn plates.

---

1 Letter from W B Thompson to J S MacLean dated 20 November 1906.

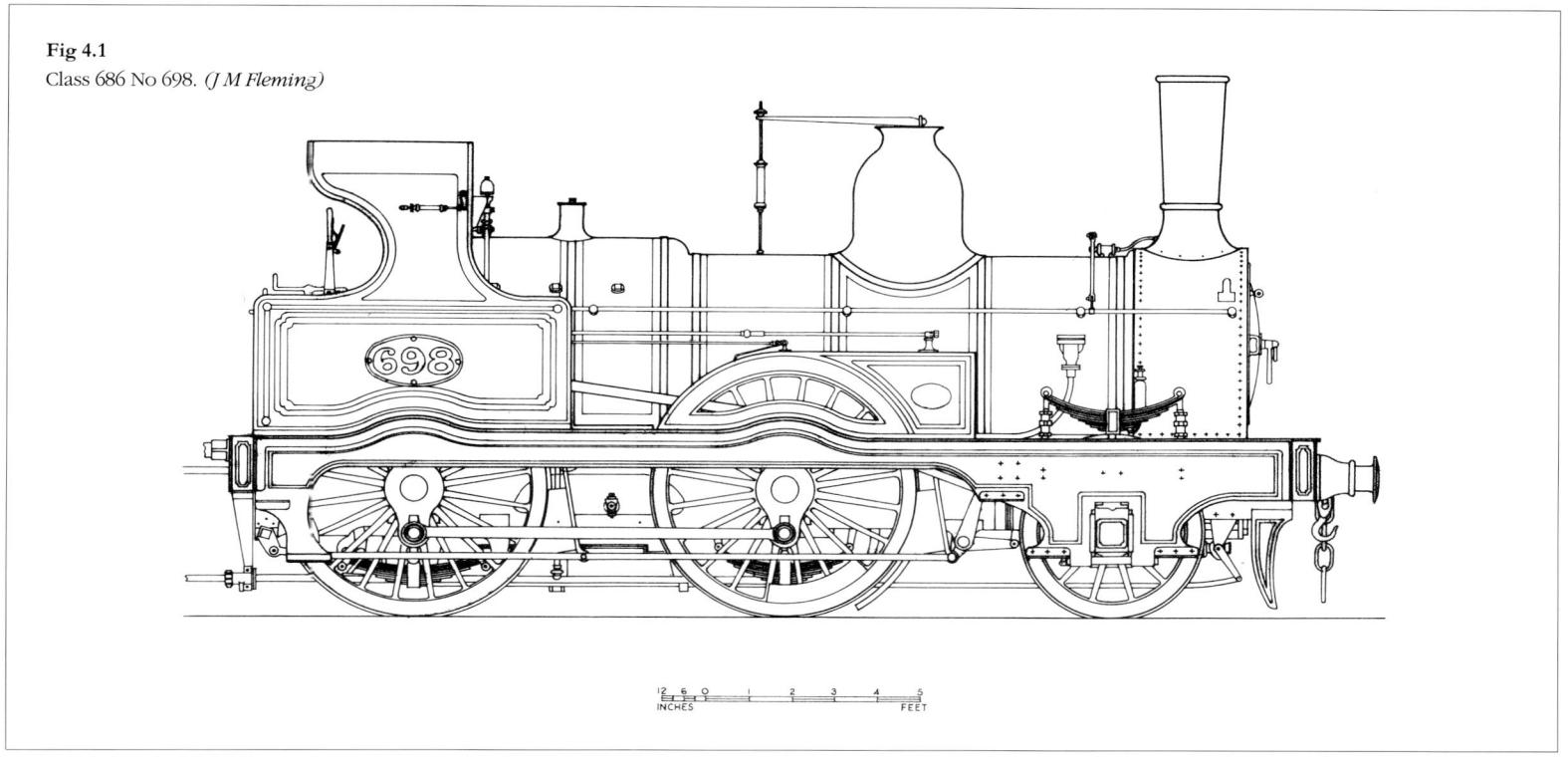

Fig 4.1
Class 686 No 698. *(J M Fleming)*

Eight more 2-4-0s – Nos 698-705 – having generally similar dimensions but a much smarter appearance were supplied by Robert Stephenson & Co in 1870-71 (Plates 4.2 and 4.3, and Fig 4.1). An obvious innovation was the provision of a cab with an arched roof and pleasingly contoured sides integrated with the footplate side-sheets, this being the prototype for Fletcher's future style of cab. The tenders had sandwich frames with iron plates on each side of wooden beams, while the tanks had a distinctive horizontal line of rivets halfway up the sides. (The 1886 designation was Class 686.)

Between March 1870 and September 1875 Gateshead Works built 34 small 2-4-0s for local and branch line traffic. The earliest examples had built-up inside frames, and cabs similar to those of Beyer, Peacock's Nos 686-697 (Plate 4.4); the design was revised in 1871 to incorporate plate frames and Fletcher's newly-adopted cab based on the Stephenson pattern (Fig 4.2). Raised firebox casings were superseded in 1873 by casings flush with the boilers and the Naylor safety valves were moved from the boiler barrels to the fireboxes; elliptical maker's plates of polished brass with incised lettering came into use at about the same date.(The 1886 designation was Class 675.)

The first of Fletcher's celebrated Class 901 express locomotives with 7ft 0in diameter coupled wheels and 17in by 24in cylinders was completed at Gateshead Works in October 1872 (Plate 4.5) and was followed by 18 more during the next four years (Fig 4.3); a further 16 were built between 1880 and 1882 with 17½in by 24in cylinders. In addition, Beyer, Peacock & Co and Neilson & Co each built ten locomotives in 1873, their dimensions being practically identical to those of Gateshead origin although differing in appearance (Plate 4.6). The boilers of Gateshead's Nos 901 and 902 and Nos 844-853 from

Plate 4.2 (below left)
Class 686. No 704, built by Robert Stephenson & Co, seen in its original condition, with the maker's plate on the sandbox.

Plate 4.3 (below right)
Class 686. No 698 as reboilered in December 1886 with a McDonnell boiler.

Plate 4.4
Class 675. No 679, built at Gateshead in November 1870 but adopting the Beyer Peacock style of cab.

Plate 4.5
Class 901. No 911, built at Gateshead in May 1875, shown in its original condition and dark green livery. The locomotive was withdrawn in 1914.

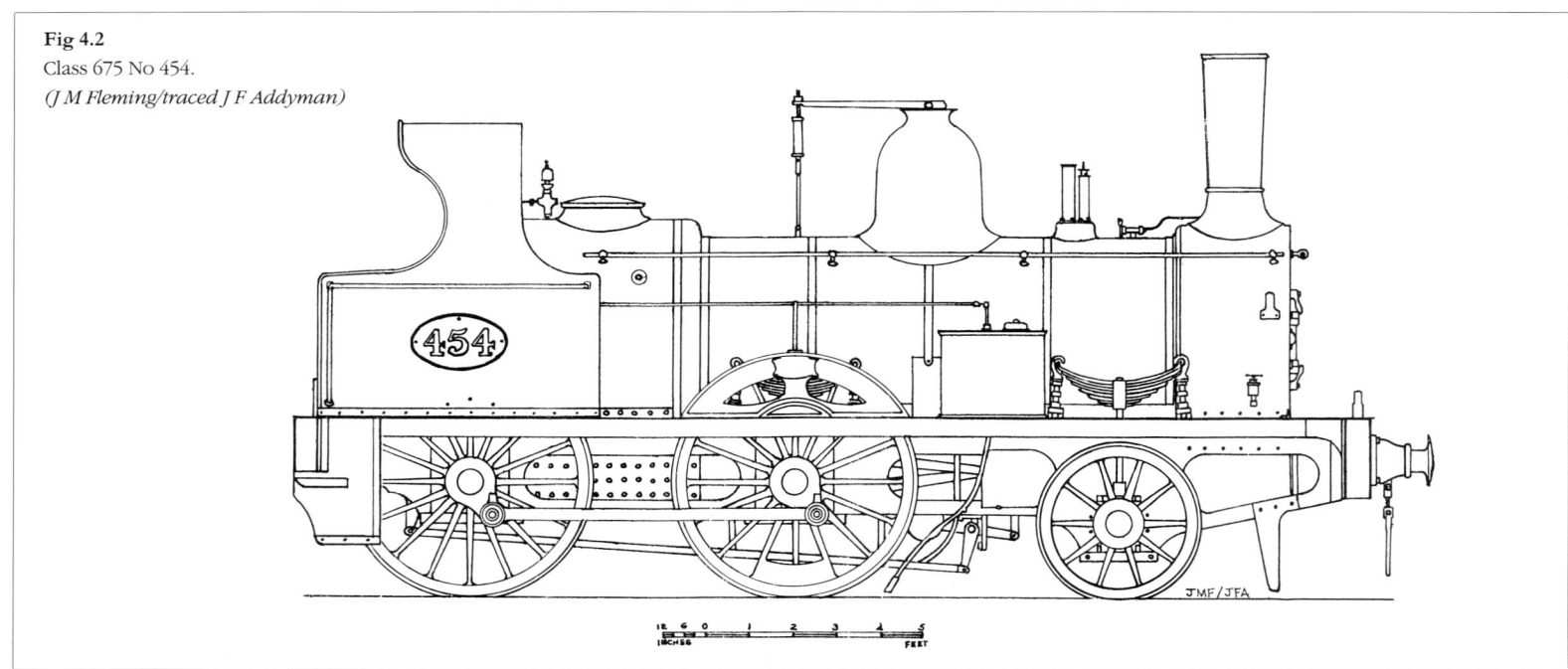

Fig 4.2
Class 675 No 454.
(J M Fleming/traced J F Addyman)

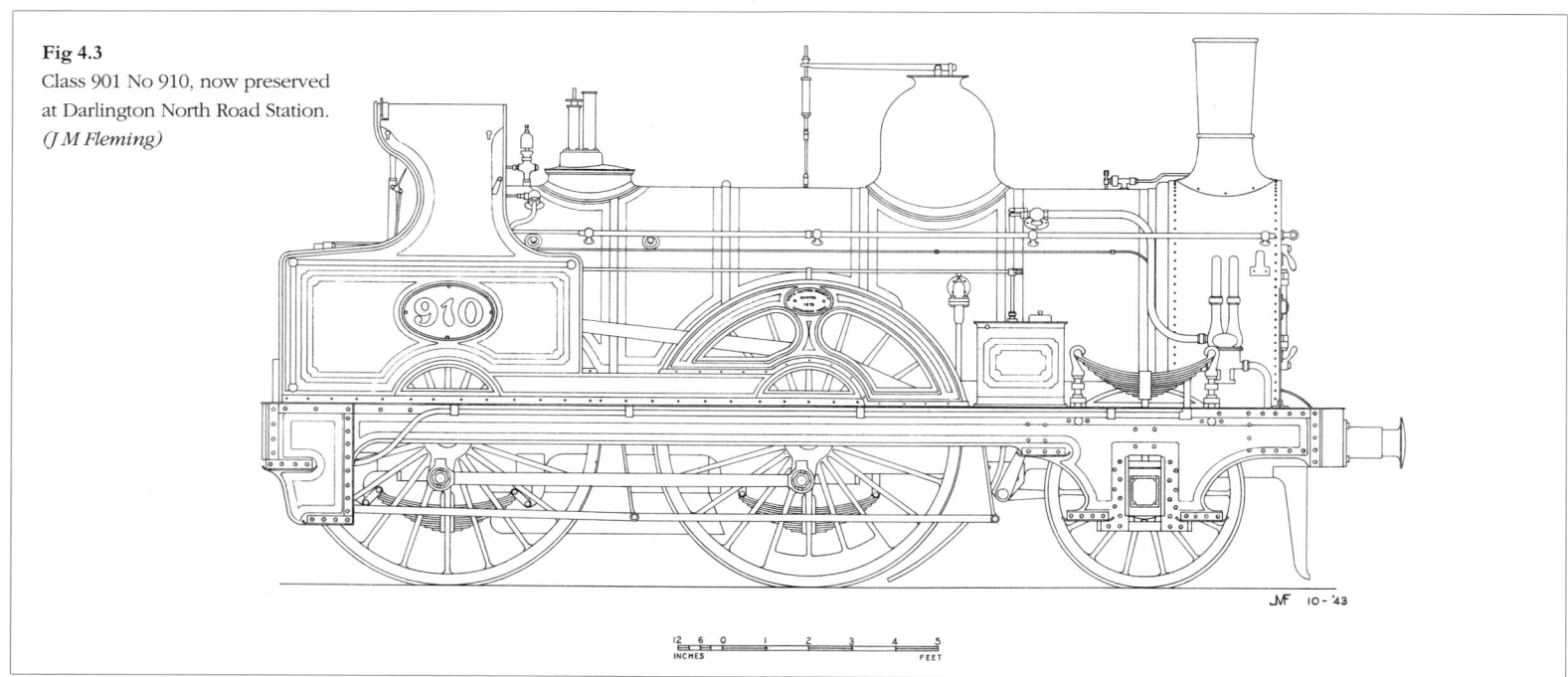

Fig 4.3
Class 901 No 910, now preserved at Darlington North Road Station.
(J M Fleming)

Plate 4.6
Class 901. No 926, seen outside Gateshead Works, was built by Neilson in December 1873 and shows minor differences. The direct-loaded safety valves and cover were an alteration. This locomotive was withdrawn in February 1914.

Plate 4.7
Class 901. No 363 was built at Gateshead in June 1880 and survived until March 1923. It is shown in the more elaborate version of the Fletcher livery originally devised for the 1875 S&D exhibition and probably adopted here for display at the 1881 Stephenson Centenary Exhibition. The pipe on the smokebox side is for the Smith vacuum brake, which had a short life on the NER, being superseded by the Westinghouse brake from the late 1870s. (Driver W Smith & Fireman J Bell)

Plate 4.8
Class 901 locomotives were generally rebuilt during their long lives. No 902, built at Gateshead in December 1872, has been transformed with a McDonnell boiler and cab as well as his brass splasher beading bearing the company's name (see Chapter 6, although the reboilering was carried out in November 1885, after McDonnell's departure. Seen here at Borough Gardens, No 902 was withdrawn in February 1913.

Beyer, Peacock initially had Naylor safety valves near the front of the barrel, but these were soon removed to the top of the firebox, which was the position adopted for the remainder down to 1876. The Naylor valves were later replaced by direct loaded valves covered by polished brass casings.

The profile of the cab sides of successive groups of Gateshead locomotives underwent a gradual change, the height of the lower panels soon being increased while the 'cut-away' of the upper part – or 'hood' – decreased in size; the cut-away was eliminated entirely in 1881 which gave the later cabs a rather clumsy appearance. Subsequently the cut-aways on early cabs were often plated-up to give better protection for the enginemen. (When No 910 was being prepared for the S&DR Centenary Celebrations in 1925, and its eventual preservation, its cab was cut down to the 1872 profile instead of to its true 1875 form.)

A solitary 2-4-0 – No 364 – with 6ft 0in diameter coupled wheels and 17in by 24in cylinders was built at Leeds Works in September 1875 (Plates 4.9 and 4.10). This was the forerunner of a class of ten locomotives – Nos 1440-1449 – built at Gateshead between October 1876 and May 1878 which, apart from the smaller wheels, was

Plate 4.9
No 364 had the unusual distinction of having been built at the NER's Leeds workshops in September 1875. It is seen here, at the east end of Leeds New Station, with a later Worsdell boiler and the tender capacity increased by the provision of coal rails.

Plate 4.10 (right)
No 364. The Leeds works plate, seen in a detail from plate 4.9.

similar to Class 901. (Plate A4.1) The smaller wheels resulted in a lowering of the running plate by six inches. Four more of these locomotives were built in 1882. All had 17in by 24in cylinders (Plate 4.11). (The 1886 designation was Class 1440.)

Passenger Tank Locomotives 1870-1876:
Edward Fletcher, Locomotive Superintendent

During the 1870s several old 2-4-0 tender locomotives were reboilered at York Works and provided with saddle tanks, bunkers and sometimes with enclosed cabs (Plate 4.12). However, after only a few years as 2-4-0STs they were rebuilt as 0-6-0STs with new frames, wheels and inside cylinders, little or nothing then surviving from the original tender locomotives.

Fletcher's 0-4-4T 'Bogie Tank Passenger' (BTP) locomotives were introduced to replace some of the elderly tender locomotives that hitherto had been used for short distance passenger services. The first to be delivered were Nos 947-958 built by Neilson & Co. in 1874, and Nos 1340-1349 and 1430-1439, built by R & W Hawthorn & Co in 1875, all having 5ft 0in diameter coupled wheels and 16in by 22in cylinders (Fig 4.4a). Ten locomotives of similar design but with 5ft 3in diameter coupled wheels were built at Gateshead Works in 1876-77. The subsequent variants of the type, built at Gateshead, Darlington and York, are described and illustrated in Chapter 5. (The 1886 designation was Class BTP.)

Goods and Mineral Locomotives 1870 - 1876:
Edward Fletcher, Locomotive Superintendent

The last 0-6-0 goods locomotives with double frames were built for the NER in 1870-73. Fifty of these were supplied by R Stephenson & Co in 1870-71, their outside frames being composed of 12in by 4in oak beams sandwiched between ½in thick iron plates; they had 5ft 0in diameter wheels and 17in by 24in cylinders. The first twenty – Nos 706-725 – had boilers with raised firebox casings and lever-operated reversing gear (Plate 4.13). The later locomotives – Nos 726-755 – had flush-topped firebox casings and Stephenson's combined lever-and-screw reversing gear. All had brass-edged sandboxes – a distinctive feature of many locomotives built by R Stephenson & Co in the 1870s.

Twenty similar locomotives – Nos 756-775 – were supplied by R & W Hawthorn & Co in 1872-73 (Fig 4.5 and Plate 4.13a). Minor external differences were apparent in the shape of the outside frames, the cabs and the leading sandboxes, the latter being much smaller than those of the Stephenson locomotives. (The 1886 designation was Class 708.)

A total of 160 six-coupled goods locomotives with inside frames were built to Fletcher's specifications by five different manufacturers between 1872 and 1876. The wheel and cylinder dimensions were the

Plate 4.11
Class 1440. No 220 is one of the later locomotives, built at Gateshead in September 1882 as No 668 and subsequently renumbered twice – first as 1476 and then as 220. It appears to be in Wilson Worsdell livery.

Plate 4.12
No 345. A York rebuild of an early 2-4-0 tender locomotive as a saddle tank, with the typical early York Works cab. Seen at Kirkbymoorside

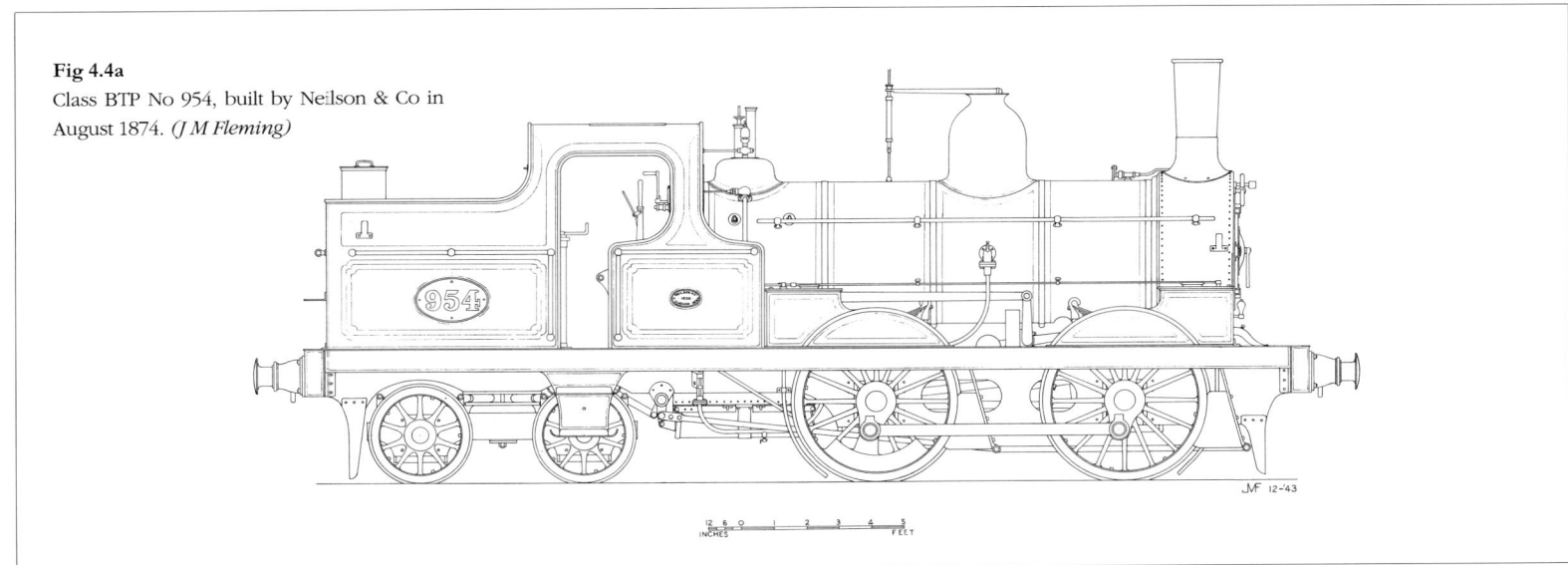

Fig 4.4a
Class BTP No 954, built by Neilson & Co in August 1874. *(J M Fleming)*

Plates 4.13 and 13a

Class 708. No 718, built by Robert Stephenson & Co in November 1870 with raised firebox and integral sandbox and splasher, whereas Hawthorn-built locomotives, such as 767 of December 1872 had a flush firebox and separate sandbox. The cab cut-out has been plated up in both cases.
(Plate 4.13 - J F Mallon collection)

Fig 4.4b

Class BTP No 273, built at Gateshead in 1878 with 5ft 6in coupled wheels instead of the 5ft 0in wheels of the original locomotives (Fig 4.4a). This variant is described in Chapter 5. Note the introduction of bogie brakes, in addition to the coupled wheel brakes, and the later style of combined cab footstep and brake cross-shaft bearing.
(J S MacLean)

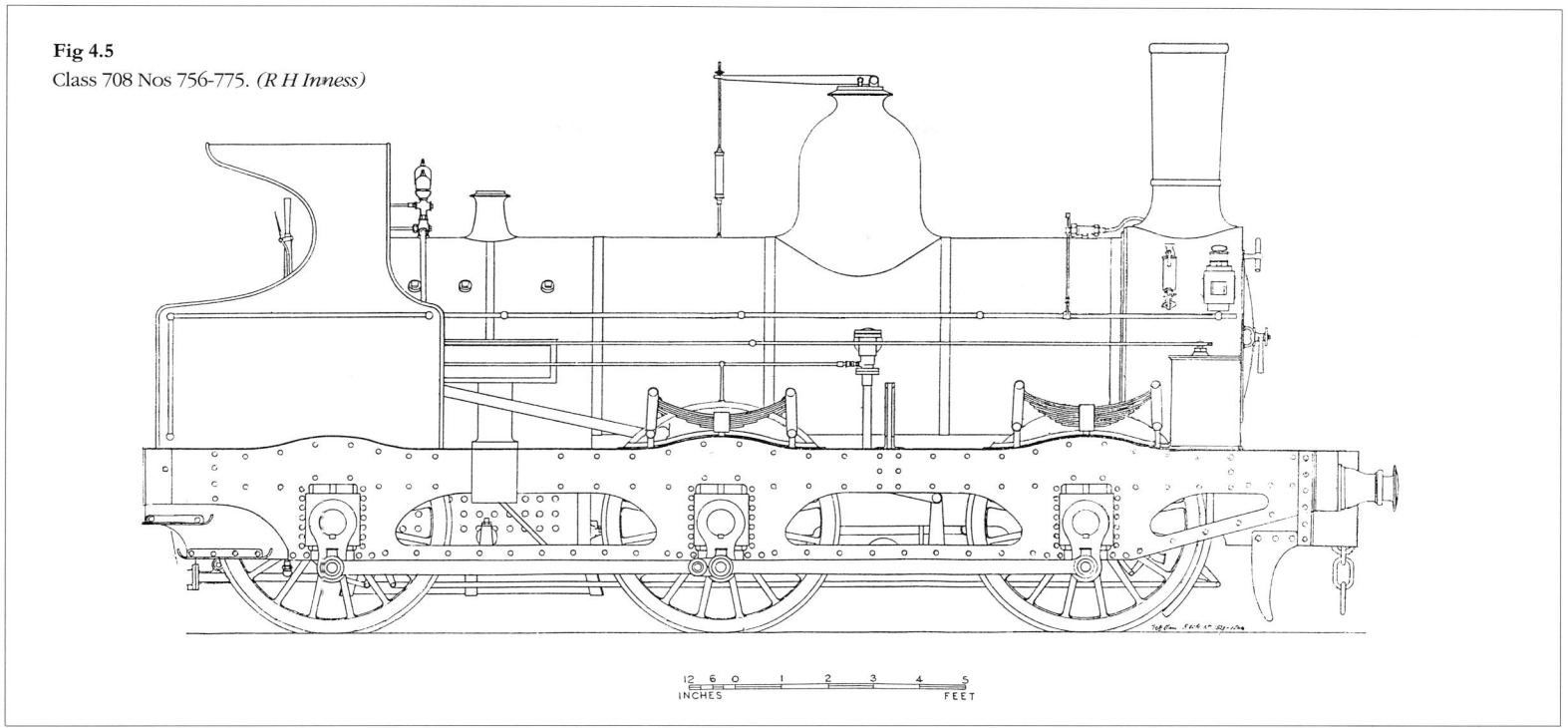

Fig 4.5
Class 708 Nos 756-775. *(R H Inness)*

same as those of the preceding outside-framed 0-6-0s but in the new design the rear of the locomotive was supported by a transverse laminated spring 3ft 9in between centres.

The first sixty, built by R Stephenson & Co in 1872-74, could be recognised by the wedge-shaped slots in the frames ahead of the leading wheels and by their large brass-edged sandboxes. Initially they were painted dark green with elaborate panelling and lining-out in the same style as on the passenger locomotives (Plate 4.14). Their numbers were: 781-799, 813-823, 879-897 and 913-923; Nos 1370-1389 from the same makers had small sandboxes. Nos 824-843 and 982-991 from R & W Hawthorn & Co in 1873-74 had open splashers, Nos 934-943 from Neilson & Co had sandboxes neatly combined with the splashers, while Nos 1410-1429 from Dubs & Co had narrow sandboxes mounted rather awkwardly on the leading splashers; all these would have had a dark green livery initially but no details of their lining-out are available. Nos 1390-1409 built by Sharp, Stewart & Co in 1875-1876, which had two small openings in each splasher, were the locomotives described by W B Thompson as "... black or practically black like the LNWR." (Plate 4.15).

Gateshead Works began turning out similar 0-6-0s in June 1875. No 174 was the second one to be completed and was shown at the S&DR Jubilee Celebrations at Darlington in September 1875 (Fig 4.6 and Plate 4.16). Small tenders with a 5ft 3in + 5ft 0in wheelbase and having a capacity of 2,000 gallons of water and three tons of coal were provided for about ten of the earliest locomotives but the majority had the more familiar type of Fletcher tender with springs above the frames. No more locomotives were built to this design after about April 1879, an improved version with frames

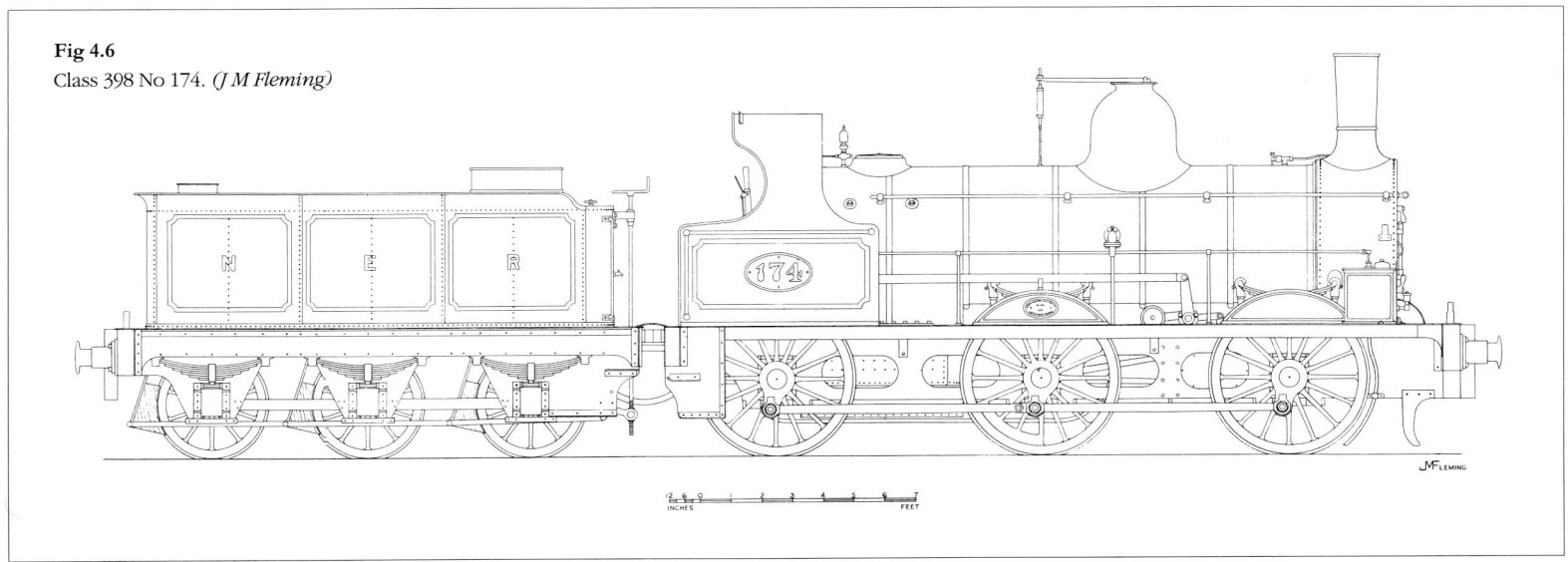

Fig 4.6
Class 398 No 174. *(J M Fleming)*

Plate 4.14
Class 398. No 813, supplied by Robert Stephenson & Co, photographed on 23 August 1881, still in original condition and with the lining on the cab front just visible. The names of the guard, fireman and driver have been helpfully added.

Plate 4.15
Class 398. No 1390, built by Sharp Stewart & Co in November 1875. Note the smokebox wingplates, which do not feature on the contemporary locomotives from Stephenson's or Gateshead Works, though adopted on some later Gateshead locomotives such as No 220 in Plate 4.11

Plate 4.16
Class 398. No 174, built at Gateshead in 1875, is shown with the smaller tender.

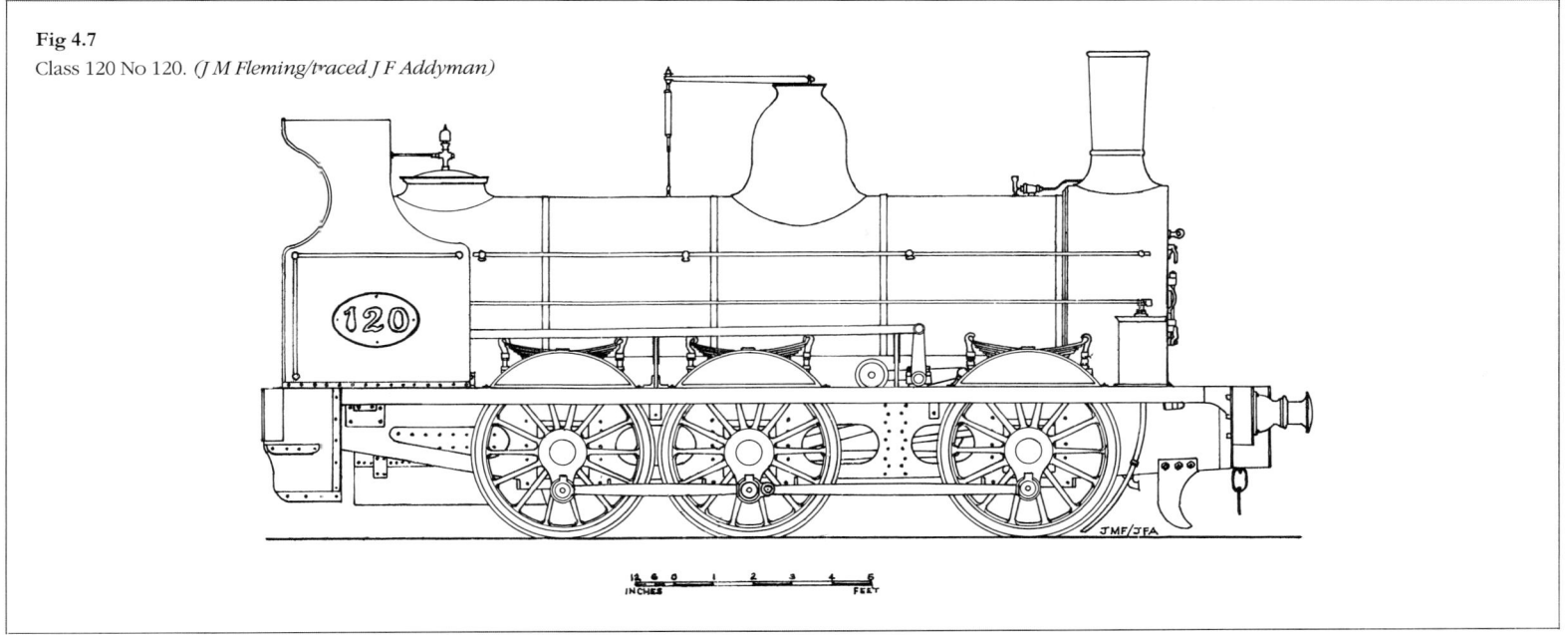

Fig 4.7
Class 120 No 120. *(J M Fleming/traced J F Addyman)*

incorporating underhung springs having made its appearance a few months earlier. The locomotives built to the modified design are described in a later section. (The 1886 designation of all the above 0-6-0s was Class 398.)

A few more long-boiler 0-6-0s with inside frames were built at Gateshead Works in the 1870s, the old design having been up-dated to include plate frames although still retaining a boiler with a raised firebox casing. Gateshead's last long-boiler 0-6-0 tender locomotive was a small 'one-off' example, No 99 built in December 1875; it had 4ft 0in diameter wheels and 15in by 22in cylinders, the wheelbase being 6ft 9½in + 4ft 3½in. (All were later included in Class 41.)

Six larger and more powerful long-boiler 0-6-0s were built at Gateshead between December 1874 and August 1875 (Fig 4.7). They had firebox casings flush with the boiler barrels, 4ft 6in diameter wheels, 17in by 24in cylinders and an 11ft 11in wheelbase, divided 7ft 2in + 4ft 9in. (The 1886 designation was Class 120.)

Shunting Locomotives 1870-1876:
Edward Fletcher, Locomotive Superintendent

The design of two short-coupled 0-6-0 saddle tanks built in 1869 (Fig 3.6) was the basis for a much-modernised version – Nos 776-780 and 854-858 – built by Black, Hawthorn & Co in 1872-73, the latter having plate frames, 4ft 2in diameter wheels and 14¾in by 22in cylinders; their narrow weatherboards gave very little protection over the footplate (Plate 4.17). Forty similar 0-6-0STs built by R Stephenson & Co in 1873-75 had 4ft 0in diameter wheels, 15in by 22in cylinders and a 6ft 6in + 5ft 2in wheelbase. These locomotives

Plate 4.17
Class 964. No 857 was built by Black, Hawthorn in July 1873 for general shunting duties and is seen in its original condition. It was renumbered 1748 in 1888.

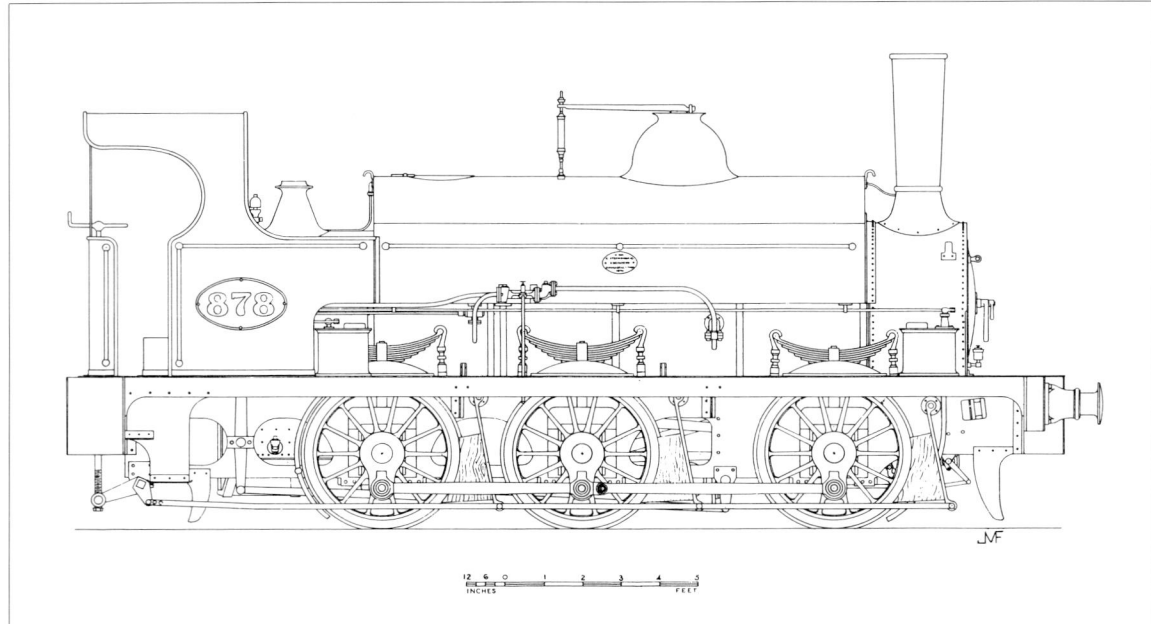

Fig 4.8
Class 964 No 878. A drawing of No 968 rebuilt as a side tank locomotive (Class 964A) appears as Fig 8.3.
(J M Fleming)

had cabs with rear weatherboards when new but the cab of at least one was soon altered to 'open back' (Fig 4.8). The numbers were 859-878 and 962-981. (The 1886 designation was Class 964.)

An entirely different design of 0-6-0ST was ordered from R & W Hawthorn & Co for the Darlington Division, twenty being delivered in 1875-76 (Fig 4.9). They had 4ft 0in diameter wheels, 15in by 22in cylinders and a 6ft 6in + 7ft 1in wheelbase (Plate 4.18). The numbers were 1350-1369. (The 1886 designation was Class 1350.)

Two powerful 0-6-0 saddle tanks with outside plate frames were built at Gateshead Works in 1875 for assisting loaded trains up the steep incline from Redheugh to Greenesfield – a line which had originally been rope-worked by a stationary steam engine. They had 4ft 0in diameter wheels and 18in by 24in cylinders, the wheelbase being 16ft 6in, equally divided: their numbers were 476 and 479 (Plate 4.19 and Fig 4.10). Two more 'incline engines' of similar design but with 'solid' plate frames – ie without the horizontal slots of the first two locomotives – were built in 1882, their numbers being 487 and 491, the former having 3ft 8in diameter wheels. No 491 was renumbered to 1474 in 1883 and 73 in 1885. (The 1886 designation was Class 476.)

From February 1868 to April 1875 six long-boiler 2-4-0 tender locomotives from a class of nine which had originated on the Newcastle & Darlington Junction Railway in 1845-6 were renewed as 0-6-0STs. Some were turned out as long-boiler, short-coupled saddle tanks but No 30 appeared from Gateshead Works in April 1873 as a conventional 0-6-0ST having a wheelbase of 14ft 8in (Fig 4.11).

Plate 4.18
Class 1350. No 1369 was brought into traffic in April 1876. It is shown at Gateshead Works, painted in a more elaborate version of the normal livery for display at the 1881 Stephenson Centenary Exhibition held at the Forth Goods Yard in Newcastle. It was later sold to Backworth Collieries, where it became No 1.

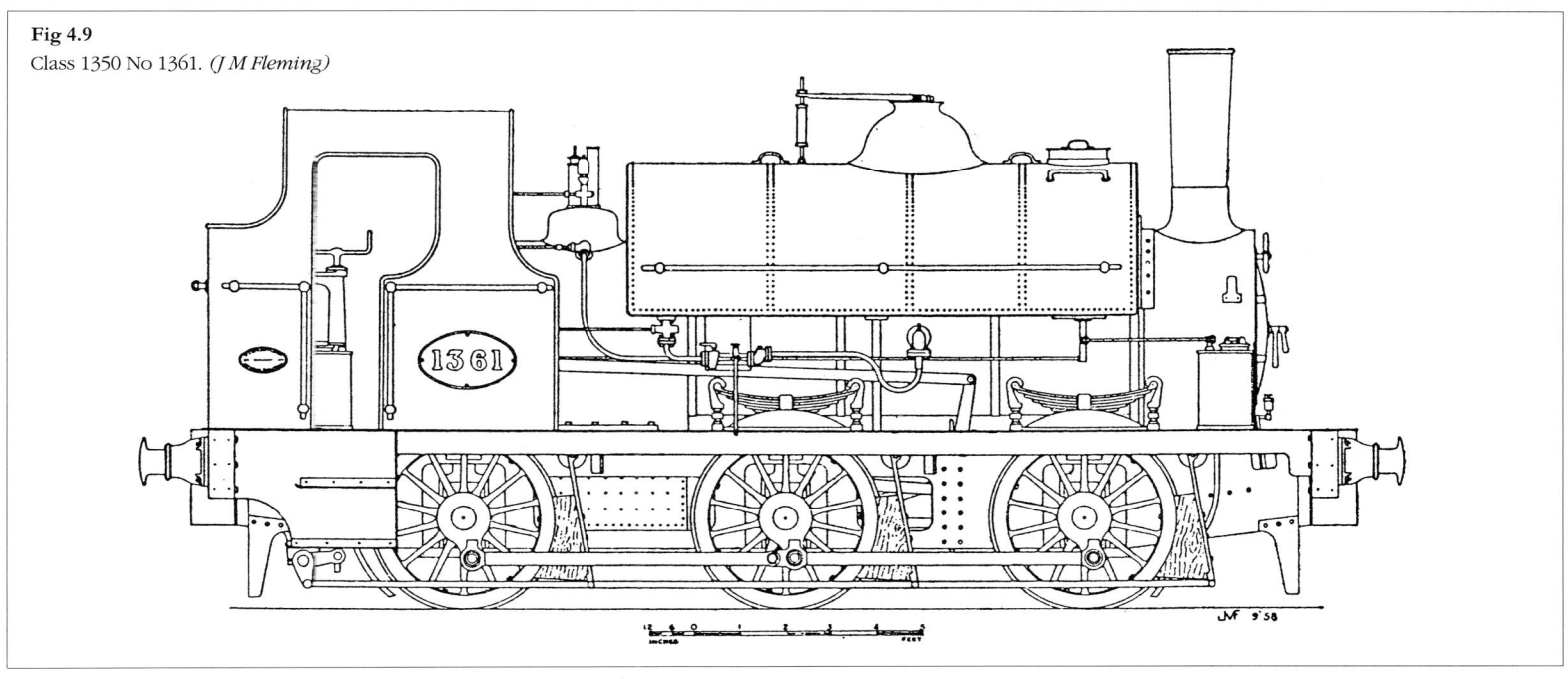

Fig 4.9
Class 1350 No 1361. *(J M Fleming)*

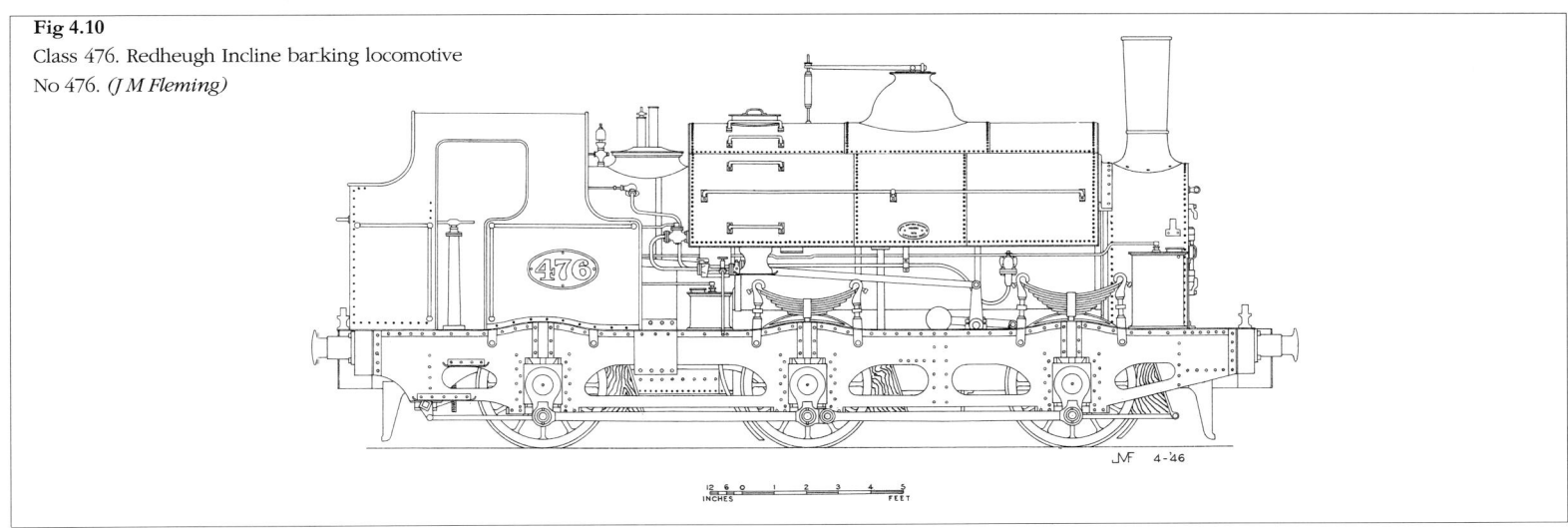

Fig 4.10
Class 476. Redheugh Incline banking locomotive No 476. *(J M Fleming)*

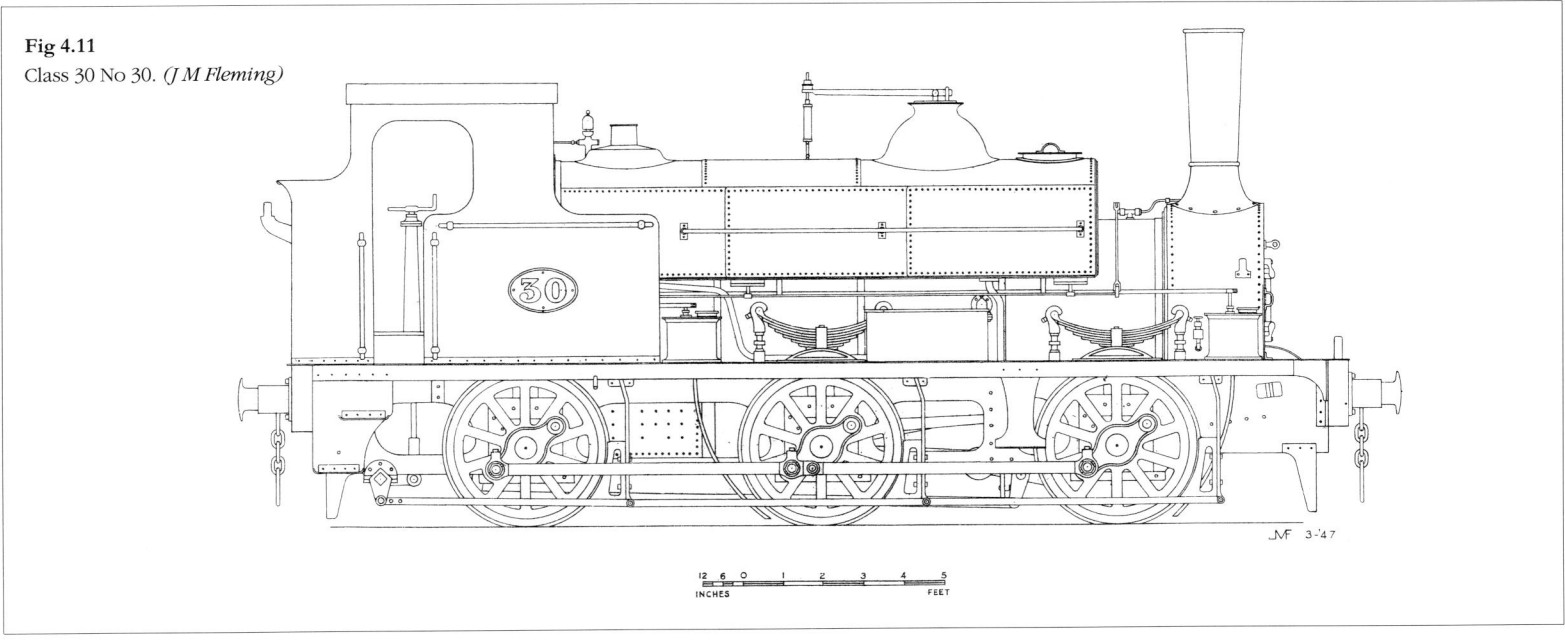

Fig 4.11
Class 30 No 30. *(J M Fleming)*

Plate 4.19
Class 476. No 479, one of the two Redheugh Incline banking engines, built in January 1875 and shown in its original condition. It was withdrawn in 1914.

Darlington Division Passenger Locomotives 1870-1876: William Bouch, Locomotive Superintendent

The first of William Bouch's 4-4-0s, No 238, was completed at North Road Works in December 1871, almost a year before Edward Fletcher's 2-4-0 No 901 appeared from Gateshead Works. Nos 239-241 followed in 1872 and Nos 1265-1270 in 1874 (Plate 4.20), the latter incorporating some detail modifications and being somewhat different in appearance. All had 7ft 1in diameter coupled wheels, with a wheelbase of only 7ft 3½in. The bogies had 3ft 7in diameter wheels, with a 6ft 6in wheelbase, and could rotate about a 'cup and ball' bearing, the total wheelbase of the locomotives being 21ft 0½in. The trailing end of Nos 238-241 was carried by a transverse open-plate spring above the axleboxes and individual volute springs below.

The boilers were 11ft long by 4ft diameter with firebox casings 5ft long by 4ft wide. The cylinders were 17in by 30in with 13in diameter solid brass piston valves instead of the usual slide valves; expansion of these heavy valves caused a great deal of trouble due to friction and the trapping of water in the valve chests. These locomotives were provided with Bouch's 'Patent Steam Retarder' a form of counter-pressure brake, which was operated by closing the regulator, putting the locomotive into 'back gear' and opening a cock to admit steam to both ends of the cylinders to cushion the movements of the pistons.

Bouch took great care to embellish his new passenger locomotives, giving them brass beadings around the cab and splashers together with burnished handrails and smokebox door hinges. The tenders

Plate 4.20
Darlington Division. No 1269, built at Darlington in 1874. These troublesome locomotives were nicknamed "Ginx's Babies" after some problematic babies featured in a contemporary political novel. They originally had curved-back tenders, illustrated by Tom Pearce on page 183 of *Locomotives of the Stockton & Darlington Railway*.

also had brass beadings along the front and top edges of the tanks to match those on the cabs, while the curving profile of the back was surely unique; the backs were later altered to the usual flat shape. The tenders held 1,600 gallons of water and 3½ tons of coke; the wheels were 4ft diameter with a 10ft wheelbase equally divided. (Incidentally, with effect from 21 October 1872 the locomotives of the Darlington Division had 1000 added to their numbers and were assimilated into the NER stock. It is possible that Nos 240 and 241, built in November and December respectively, were turned out with the 1000 already added.)

Nos 1265-1270, built in May-November 1874, had a different design of splashers over the coupling rods and driving wheels, a shorter dome with the safety valve levers above the rim, and independent underhung springs for the trailing axleboxes. No 1270 was exhibited at the S&DR Jubilee celebrations in Darlington in September 1875.

After these 4-4-0s came under Mr Fletcher's control they were rebuilt as 2-4-0s (see Plate 5.2). (The 1886 designations, after rebuilding, were Class 1238 and Class 1265.)

Plate 4.21
Darlington Division, Class 1068. No 1035.
(Ken Hoole collection)

Six 2-4-0 passenger locomotives built between August 1875 and July 1876 were much more successful than the 4-4-0s. They had 6ft 0½in diameter coupled wheels and a 4ft 8in diameter leading pair, with a 16ft 6in wheelbase divided 8ft 6in + 8ft 0in: the cylinders were 17in by 26in. The boilers were 11ft 2in long by 4ft 0in diameter with firebox casings 4ft 10in long. Even in their original condition they

Fig 4.12
Class 1001 No 1275. Darlington Division. The shaped buffer beam was provided to accommodate a second, lower set of sprung buffers originally included for use with chaldron wagons (see *North Eastern Record Volume 2* - Chapter 9).*(L Ward)*

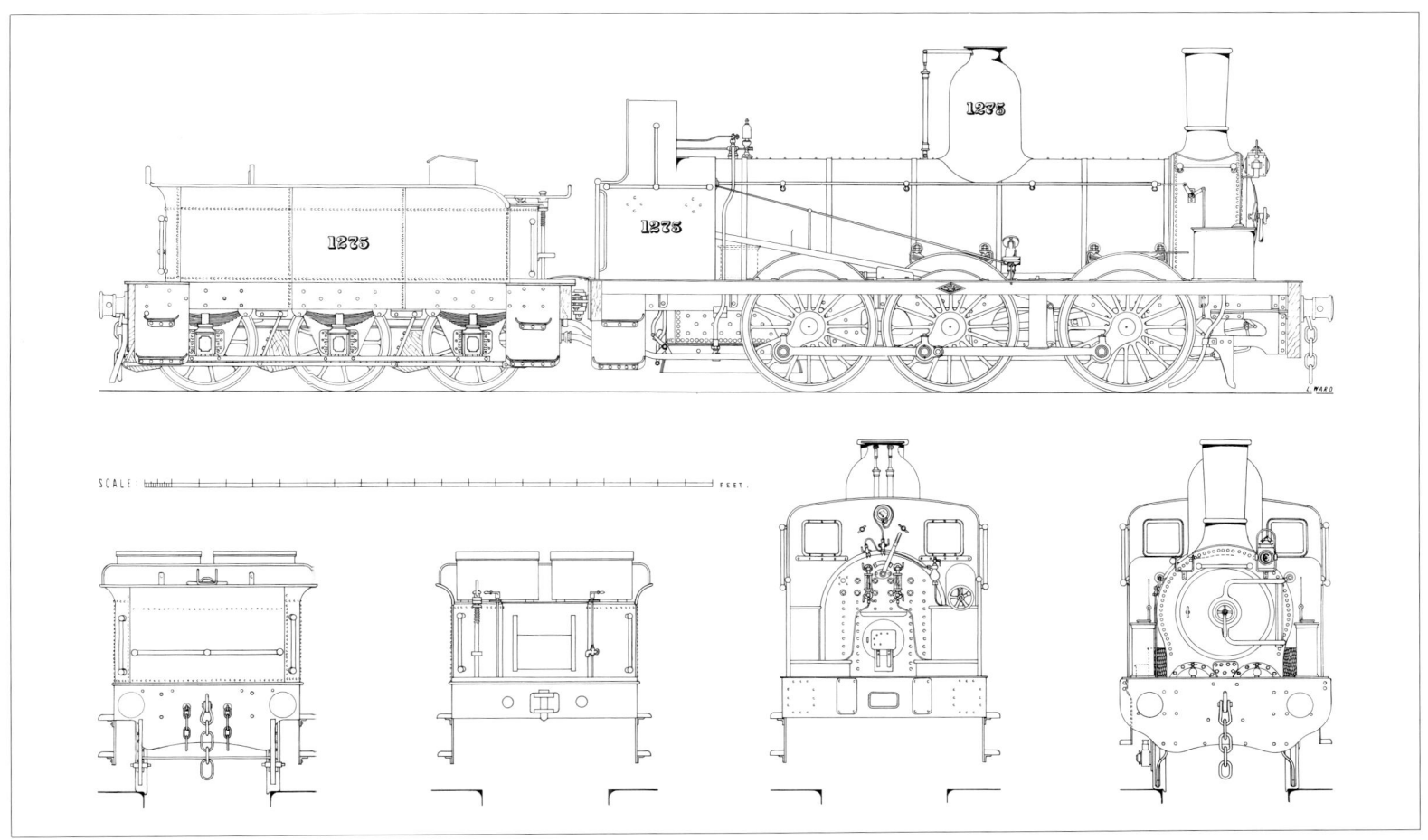

Plate 4.22
Darlington Division, Class 1001. No 1289, built by Avonside Engine Co in March 1875. Dumb buffers, for use with chaldron wagons, can be seen below the buffer beam. *(J F Mallon collection)*

were not all alike in appearance, there being at least two varieties of splashers, cabs and domes (Plate 4.21). No 1068 had smaller leading wheels than the others and a fanwise arrangement of slots in the driving wheel splashers. It participated in the S&DR Jubilee Celebrations in 1875 with its tender lettered N E R for the occasion. Initially the locomotives were provided with 2,400 gallon tenders of Bouch's design, but these were exchanged quite soon for Fletcher's 2,500 gallon pattern. (The 1886 designation was Class 1068.)

Darlington Division Goods and Mineral Locomotives 1870-1876: William Bouch, Locomotive Superintendent

Between August 1870 and December 1873 another 24 of the Darlington 'standard' long-boiler mineral locomotives, by that time fitted with Bouch's steam retarder, were built at North Road Works; the first six were replacements for old locomotives, the others – Nos 231-237, 242-245 and 250-256 – were additions to the Darlington stock. Twelve similar locomotives were also supplied by Gilkes, Wilson & Co between December 1872 and May 1875, numbered 246-249 and 257-264. Those locomotives built after October 1872, i.e. Nos 242-264, would probably have had 1000 added to their numbers before being taken into stock.

The last long-boiler 0-6-0s built by contractors were Nos 1271-1280 from Dubs & Co (Fig 4.12) in May and June 1874, and Nos 1281-1290 from the Avonside Engine Co. between November 1874 and March 1875. The wheels were 5ft 0½in diameter, the cylinders 17in by 26in and the wheelbase 11ft 10in, divided 6ft 7in + 5ft 3in. The rear end of the Dubs' locomotives was supported in Bouch's usual manner by a large transverse plate spring above the trailing axleboxes. The Avonside locomotives had independent underhung springs for the trailing axleboxes, their rear hangers being secured to the sides of the firebox. (Plate 4.22)

Construction of long-boiler 0-6-0s at North Road Works came to an end in 1875 with Nos 1067 and 1079 built in January and No 1032 in February and No 1027 in March. They had 17in by 26in cylinders, 5ft 0½in diameter wheels, an 11ft 8in wheelbase divided 6ft 5in + 5ft 3in and independent underhung springs for the trailing axleboxes. (The 1886 designation of these long-boiler locomotives was Class 1001).

A new design of 0-6-0 goods locomotive with a longer wheelbase and correspondingly shorter boiler was produced in 1875, eight of which were built at North Road Works. (Plate 4.23) They had 5ft 0½in diameter wheels and 17in by 26in cylinders, but there were variations in the wheelbase. The first two, Nos 1038/45 had a 15ft 7in wheelbase, divided 7ft 9½in + 7ft 9½in: on Nos 1052 and 1921 the wheelbase was 15ft 10in, divided 8ft 0½in + 7ft 9½in and on Nos 1063/69/70/74 it was 16ft 3in, divided 8ft 3in + 8ft 0c. (Dimesions noted from working drawings by R H Inness.) The boiler barrels were 11ft 2in long by 4ft 0¾in diamiter with firebox casing 4ft 10in long and 4ft 0in wide. The locomotives had No 8 Friedmann injectors instead of the usual pumps, Bouch's reversing screw and patent steam retarder. (These locomotives were later included in Class 398.)

William Bouch died on 19 January 1876, but his influence was still apparent in the design of four powerful 0-6-0 side tank locomotives

Plate 4.23
Darlington Division, Class 398. No 1070, built at Darlington in 1875. Conspicuous on the middle splasher is the works plate, reading:

> North Road Engine Works
> December
> 1875
> No. 117
> Darlington

(Ken Hoole collection)

built at North Road Works between September and December of that year and intended for working on the Skinningrove zigzag incline. The dimensions of the cylinders were 18in by 26in, the wheels were 4ft 8½in diameter and the wheelbase 16ft 6in, divided 8ft 6in + 8ft 0in. The motion, axles and axleboxes were practically interchangeable with those of the preceding 2-4-0s; the boilers were also similar, with the exception that the firebox casings were flat-topped.

The former S&D system came fully under control of the North Eastern Board at York on 31 December 1876 and in the following year the Darlington Works began to build locomotives of Edward Fletcher's design

The Blyth & Tyne Railway (B&TR)

The B&TR, which was absorbed by the NER on 1 July 1874, had its origin in the Seghill Railway, a mineral line opened in 1840 to convey coals from Seghill Colliery to staiths on the River Tyne near Percy Main. By the time of the amalgamation the B&TR had developed a network of lines in south-east Northumberland and carried passengers in addition to goods and minerals. The company's workshops at Percy

Plate 4.24
Blyth & Tyne Railway. One of the 2-4-0 passenger locomotives, with its distinctive B&T cab, reboilered by Fletcher and seen as NER No 1328. *(K L Taylor collection)*

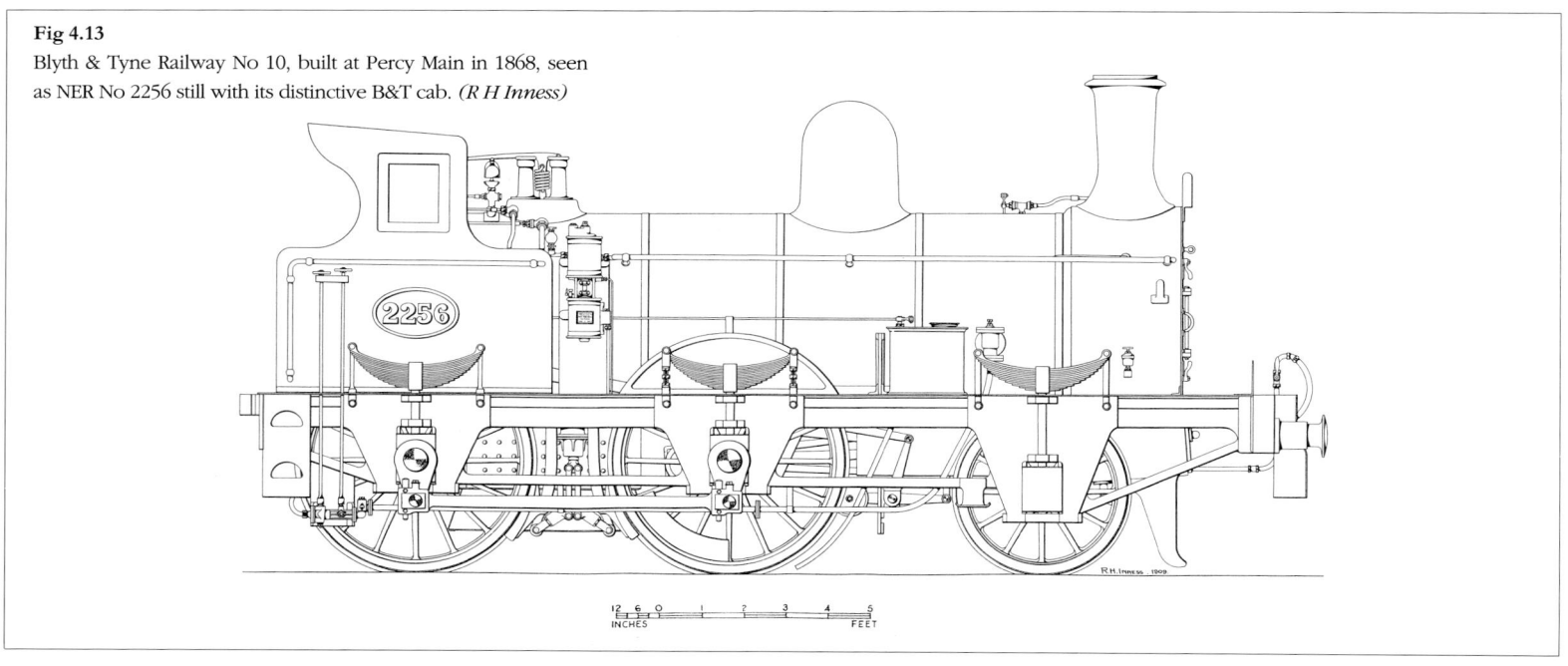

Fig 4.13
Blyth & Tyne Railway No 10, built at Percy Main in 1868, seen as NER No 2256 still with its distinctive B&T cab. *(R H Inness)*

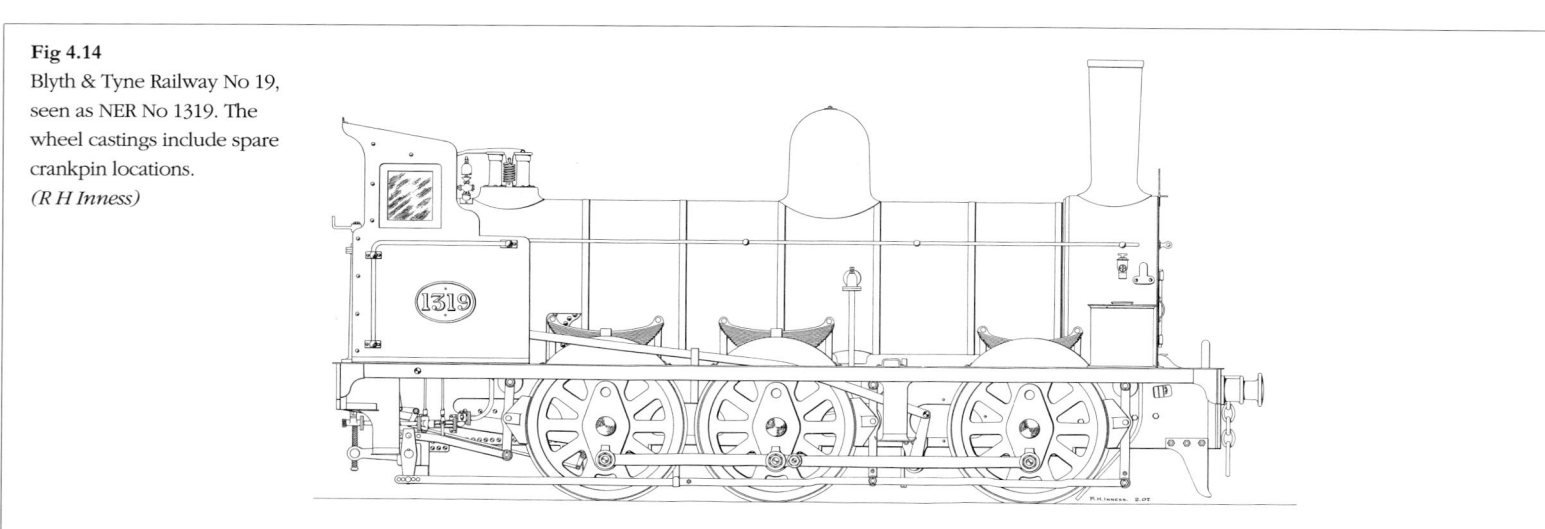

Fig 4.14
Blyth & Tyne Railway No 19, seen as NER No 1319. The wheel castings include spare crankpin locations.
(R H Inness)

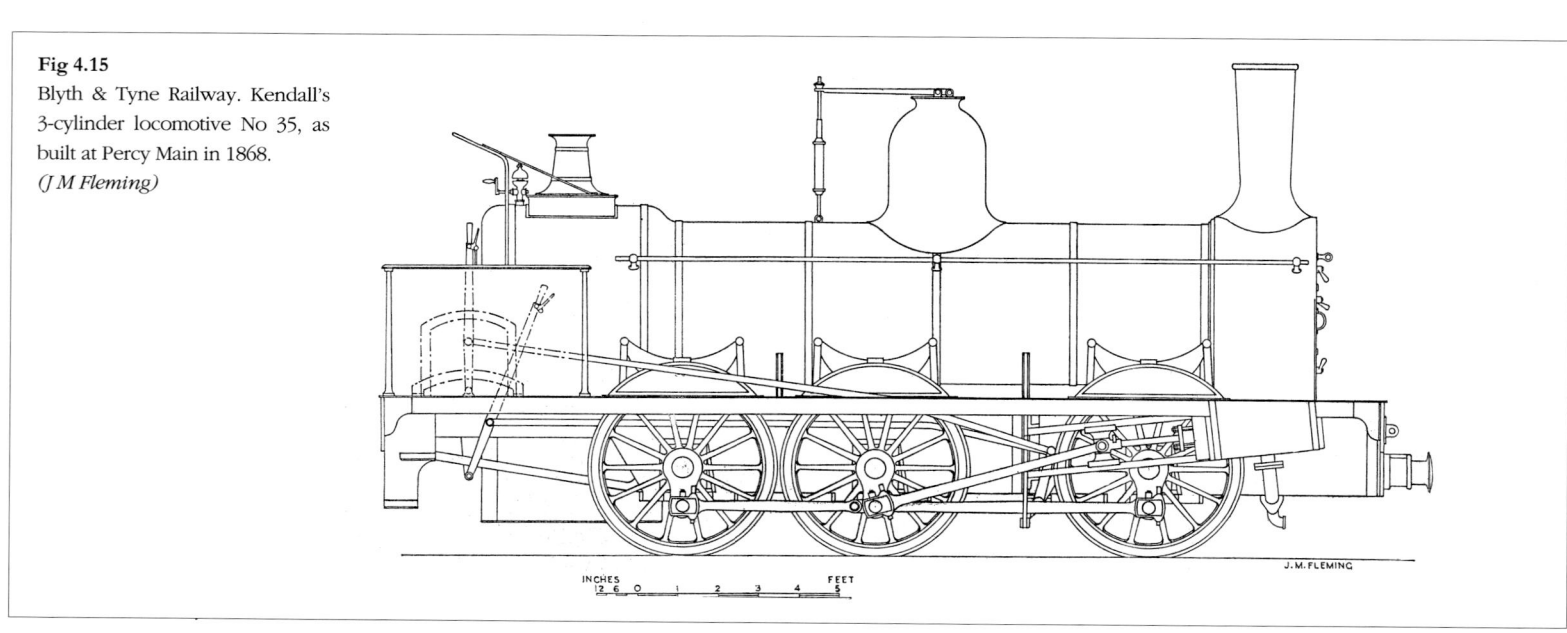

Fig 4.15
Blyth & Tyne Railway. Kendall's 3-cylinder locomotive No 35, as built at Percy Main in 1868.
(J M Fleming)

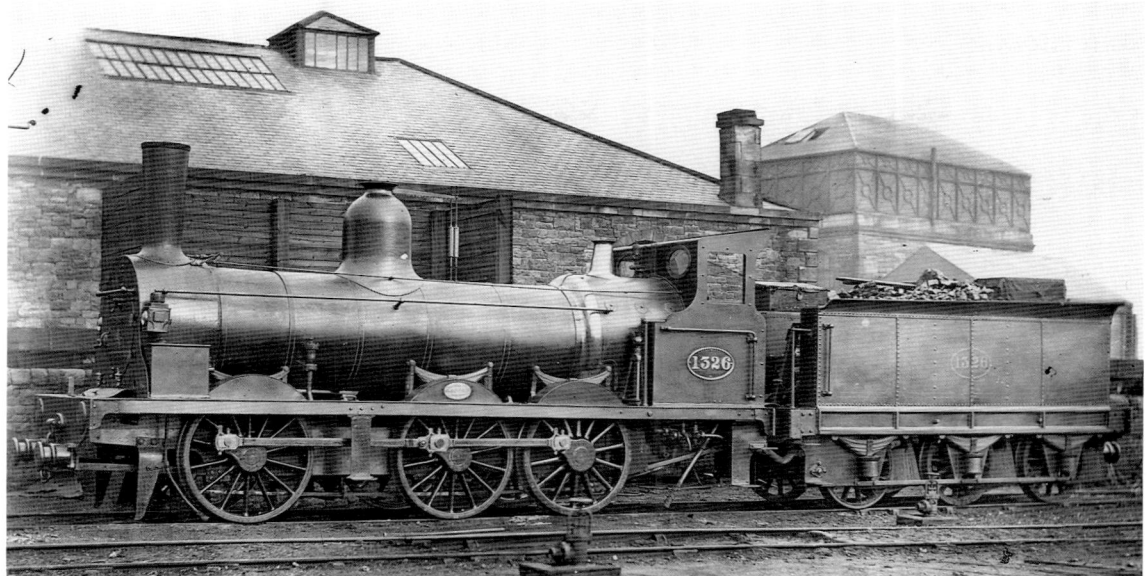

Plate 4.25
Blyth & Tyne Railway. 0-6-0 mineral locomotive No 26, built by the B&TR in March 1865 at Percy Main, where it is seen bearing its NER number 1326. It had 4ft 6in wheels and 16in by 24in cylinders. Though seen here with only the lower buffers required for handling chaldron wagons, it may once have had normal buffers as well. Renumbered 1786 in April 1889 and 1739 in January 1894, it was cut up at Percy Main in June 1897.
(Real Photographs)

Main were equipped to build locomotives, carriages and wagons, the Locomotive Superintendent being William Kendall.

In 1874 the B&TR passenger locomotives were all 2-4-0s with 5ft 6in diameter wheels and 16in by 24in cylinders, some having outside frames and some with mixed frames (Plate 4.24 and Fig 4.13). Most of the mineral locomotives were short-coupled 0-6-0s with inside frames, wheel diameters from 4ft 6in to 4ft 9in and 16in by 24in cylinders (Plate 4.25 and Fig 4.14). An interesting but obscure locomotive was Kendall's patent 3-cylinder 0-6-0 No 35, built at Percy Main in 1868. The general arrangement drawings show one large inside cylinder and two smaller outside cylinders projecting from the frames like hammer heads; the connecting and coupling rods on either side worked in phase, the crank pins being in line with each other instead of at a right angle! (Fig 4.15). After rebuilding with two inside cylinders, No 35 remained at work until 1900, being then NER No 1335 (Plate 4.26).

From about 1868 William Kendall began to add sides to the existing turned-back weatherboards and form the distinctive B&TR cabs. The sides being lined with vertical wooden boards in which grooves were cut to allow the window frames to be lowered when required. The former B&TR 0-6-0 No 8 was sold in 1907 – as NER No 1719 – to the Seaham Harbour Dock Co, where it worked under the name *Ajax* until 1926. The cab was then transferred to another ex-NER 0-6-0 - Class 120 No 125. It has outlasted both locomotives, and is at the Stephenson Railway Museum, near Percy Main.

The B&TR handed over 39 locomotives at the amalgamation, these being renumbered 1301-1339 by the NER.

Plate 4.26
Blyth & Tyne Railway. Kendall's 3-cylinder locomotive, No 35, altered to 2 cylinders and seen as NER No 1335.

Chapter 5

Locomotive Development 1877 – 1882
Edward Fletcher, Locomotive Superintendent

Passenger Locomotives 1877-1882

Six more of Fletcher's large 2-4-0s with 7ft diameter coupled wheels (Class 901) were built at Gateshead in 1880, six in 1881 and a final four in 1882. All these had 17½in by 24in cylinders, longer springs for the coupled wheels with india-rubber cushions incorporated in the hangers, and direct-loaded safety valves under polished brass covers. The last few locomotives had butt-jointed boilers with shorter domes. By 1878 the locomotive livery had already reverted to the pre-1870 bright 'Saxony', or grass green colour with Brunswick green borders but now with dark claret-coloured outside frames. Possibly the inside frames may still have been painted Indian red.

Gateshead Works also built four more 2-4-0s with 6ft diameter coupled wheels in 1882 (Class 1440). These had 17½in by 24in cylinders and other modifications similar to those mentioned above (Plate 4.11).

During the years 1877-1882 the North Road Works continued to have considerable freedom of action, several new types of locomotive designed and built there being unlike those currently being built at Gateshead.

The first amongst these were four 2-4-0 passenger locomotives with inside frames built at North Road in 1877, the diameter of the coupled wheels being 6ft 6in, the leading wheels 4ft 8in and the wheelbase 16ft 6in, divided 8ft 11in + 7ft 7in; the cylinders were 17in by 26in, the boilers 11ft long by 4ft diameter and the fireboxes 5ft long. Nos 11 and 1166, built in October 1877, were rebuilt in 1885 with 17in by 24in cylinders and with outside frames and bearings for the leading axle (Plate 5.1); Nos 1100 and 1114 retained the 17in by 26in cylinders and inside frames. All four were fitted with longer boilers and fireboxes in 1889/90 which involved reducing the length of the cabs. Nos 1100 and 1166 were later fitted with standard 'Class 901' boilers and the wheelbase was altered to 8ft + 8ft 6in. No 1166 was given a larger, angular cab in 1906, of the type favoured by Ramsay Kendal, the Works Manager at North Road. The original livery of these locomotives would have been similar to that of 2-4-0 No 1035 (Plate 4.21). (The 1886 designation was Class 11.)

Plate 5.1
Class 11. No 1166, built at Darlington in 1877 and seen in largely original condition, though it has acquired a McDonnell chimney, and still distinctly a Darlington Section design.

Colour Plate 1
York, Newcastle & Berwick Railway No 144, built by Robert Stephenson & Co in July 1847 and withdrawn about 1867. The livery depicted is Saxony green with Brunswick green borders; black panel bands and boiler lagging belts edged with a yellow line; inside frames Indian red; vermilion buffer beams; black smokebox and running plate.

(J S MacLean collection)

Colour Plate 2
An 0-6-0 taking a coal train across the temporary High Level Bridge at Newcastle during 1848-9, from a drawing prepared ten years later under the direction of Robert Hodgson who was Resident Engineer for the bridge construction. The livery is Fletcher's early NER (and late YN&BR) one of Saxony green, though shown without Brunswick green borders; buffer beam, number plate and its replica on the tender are vermilion; outside frames a dark maroon.

(Reproduced by kind permission of the Institution of Civil Engineers)

Colour Plate 3
Sample panel of NER green, with boiler lining, prepared by the painter who restored the *Tennant* No 1463 for preservation in 1925. He noted the following details: buffer beams, red edged 1in black with $3/16$in white; wheels, green, outer edge of tyre black with $3/16$in white line, centres black with red edge; footplate angle iron and steps, black with $3/16$in red line. R H Inness noted that NER green was mixed from equal parts of Prussian blue with middle Chrome yellow.

(Private collection)

Colour Plate 4
T W Worsdell introduced this armorial device into the NER locomotive livery. His brother subsequently made use of the larger heraldic device illustrated in Volume 1. The blue background is highly inappropriate.

(G F Hartley collection)

Colour Plate 5
Class R No 2015 depicted in its original livery in this July 1907 *Railway Magazine* colour plate. It is based on an official photograph in grey, which has misled the colourist into depicting the upper part of the frames and smokebox saddle as green; in practice these were painted black. Those locomotives rebuilt with the deeper Raven frames, similar to those used on Class R1, had the frames painted green above the footplate and lined out in the same way as is seen on the colour plate of Class R1.

Colour Plate 6
Class R1, No 1238 shown in its original livery in the F Moore postcard based on an official photograph of the locomotive painted in lined grey.
(Private collection)

Colour Plate 7
Class F, No 115, depicted by F Moore in its original T W Worsdell livery, in this postcard based on an official photograph in lined grey.
(National Railway Museum)

Colour Plate 8
Class X, No 1352 in its working livery of black, lined with vermilion; the grey tones of the boiler and cylinders arise from the artist's attempt to depict light reflected off the curved black surfaces. It is based on an official photograph in lined grey but this was printed in reverse and so what we see is actually the left-hand side of the locomotive. The only errors this introduces are the reversed smokebox door hinges and the absence of the variable blast pipe control rod which emerged from behind the side tank on the right-hand side of the locomotive. This plate was published in the *Railway Magazine* for May 1910.

Colour Plate 9
Class Z, No 733 in its original livery. In this March 1913 *Railway Magazine* plate the colourist has omitted the humps projecting from the coupling rod splasher in order to clear the crank pins, because they were indistinct on the original official photograph.

Colour Plate 10
Class I, No 1329 shown in its original compound form and in T W Worsdell livery.
(Painting by V Welch, private collection)

Colour Plate 11
Class Q, No 1877 depicted in its original livery, though the guard irons at the front should be black instead of red. *(Painting by V Welch, private collection)*

Colour Plate 12
Class 3CC, No 1619, shown in its original livery.
(Painting by V Welch, Darlington Railway Centre & Museum)

Colour Plate 13
Class T, No 2122, shown in its short-lived green livery, which was superseded by black following a directive of June 1904. One of the first ten to be built, it is shown with the early addition of a leading sandbox integrated with the front splasher (compare with Plate 8.22).
(Painting by V Welch, private collection)

Photographs of V Welch paintings by Ron Prattley

Colour Plate 14
A contemporary model based on one of Fletcher's Class 901 express locomotives. The number, 1882, appears to refer to the year in which it was made, which was also the year of Fletcher's retirement, The tender is shown lined into two panels rather than the customary three, though this treatment can also be seen in Plate 4.14. Unlike contemporary photographs it gives a good view of the lining out of the cab roof. The model was made by Mr R Cairns of Darlington and presented to the Bowes Museum by Mr H G T Barningham.
(Reproduced by courtesy of the Bowes Museum, Barnard Castle)

Colour Plate 15
This York, Newcastle & Berwick Railway heraldic device was used on the company's carriages, though never appearing on it locomotives.
(G F Hartley collection)

Colour Plate 16
It is unlikely that heraldry ever appeared on Leeds Northern Railway locomotives, but this device was employed on the company's carriages. Along with the YNBR left and the York & North Midland Railway device (lifted from the Arms of the City of York), it makes up the NER device shown in Colour Plate 4.
(G F Hartley collection)

The North Eastern Railway in Preservation

Colour Plate 17
Class M, No 1621, as preserved in Wilson Worsdell livery at the National Railway Museum, and seen at York Station in 1977.

Plate 5.2
Class 1265. No 1268, a Bouch 4-4-0, built in September 1874, seen after rebuilding as a 2-4-0 in 1881. *(Ken Hoole collection)*

Between March 1879 and September 1882 Bouch's troublesome 4-4-0s were all rebuilt as 2-4-0s with 17in by 26in inside cylinders and 4ft 8in diameter leading wheels. There were variations in the wheelbase of individual locomotives after the alterations: Nos 1238-1240 had 9ft 2¼in + 7ft 3½in, Nos 1265-1270 had 9ft 3in + 7ft 3½in (Plate 5.2) and No 1241, which had a longer firebox than the others, 8ft 9¾in +7ft 8½in. Modifications to accommodate the larger fireboxes of the later standard boilers included the provision of outside frames and bearings for the leading wheels. The first locomotives to have standard boilers were Nos 1265 and 1270 in 1894, their appearance then being similar to the Class 1463 *Tennant* 2-4-0s of 1885. There was rather less uniformity in the lineaments of the subsequent rebuilds. (The 1886 designations were Classes 1238 and 1265.)

The 7ft diameter coupled wheels off No 1162, one of the 4-4-0s built by R Stephenson & Co in 1862, were incorporated in a 2-4-0 built in 1880. Although similar to the rebuilt locomotives described above it was given a separate designation as Class 1162.

Four 2-4-0 express passenger locomotives of a new design were built at North Road Works in June 1882, their numbers being 40, 58, 1099 and 1101 (Plate 5.3). They had 6ft 6in diameter coupled wheels and a

Plate 5.3
Class 40. No 40, built at Darlington in June 1882 and bearing a typical Central Division cab. It has the late Fletcher Darlington livery with concave corners to the panels, three panels on the tender side and the cab hood and front lined to match. The Darlington numberplate may have a black background; it is surrounded by a fine white line and the numbers are burnished.

Plate 5.4
Class BTP. No 1000 was built at Gateshead in June 1880 and is shown in its original condition. Like many others of its class it was rebuilt as a Class 290 0-6-0T. This was done at York in 1900 and the locomotive then survived until May 1951. The painting of the initials NER on the footplate valance was a very unusual feature.

4ft 8in diameter leading pair, with a 16ft 6in wheelbase divided 8ft 7in + 7ft 11in, the cylinders being 17in by 24in. The boilers were 11ft 2in long by 4ft 2in diameter with firebox casings 5ft 0in long. These locomotives were a development of the four 2-4-0s built in 1877, with some modifications derived from Gateshead's practice, eg cylinders with 24in stroke and outside frames and bearings for the leading axles. Their external appearance was enhanced by the smooth finish of the smokebox – with countersunk riveted heads – and by the latest improved shapes of the boiler mountings and cab. (The 1886 designation was Class 40.)

Passenger Tank Locomotives 1877-1882

Following on from the earlier BTPs supplied by Neilson & Co in 1874 and Hawthorn's in 1875 the first series of ten 0-4-4Ts built at Gateshead Works was completed between June 1876 and July 1877; another five locomotives were built between October 1878 and January 1879. The coupled wheels were 5ft 3in diameter, with a 7ft 8in wheelbase, the bogie wheels being 3ft 0in diameter with a 5ft wheelbase; the total wheelbase was 21ft 8in. The boilers were 10ft 6in long by 4ft 2in diameter.

No more 0-4-4Ts with 5ft 3in dia wheels were built after January 1879 but meanwhile Gateshead Works had started to turn out locomotives with 5ft 6in diameter coupled wheels, 33 of these being

Plate 5.5
Class BTP. A Darlington product, No 1115, seen at Loftus about 1895 in original condition. It was completed in June 1877 and rebuilt as Class 290 in 1901; withdrawn November 1959. Compared with Gateshead locomotives it has a broader chimney and more slender, taller dome. Note the rear sandboxes.

Fig 5.1
Class 398 No 634. *(J M Fleming)*

built between September 1877 and October 1882 (Plate 5.4 & Fig 4.4b). Six more were built in 1883-4 after Fletcher's retirement.

Ten 0-4-4Ts with 5ft 0in coupled wheels were also built at Gateshead Works between March and December 1879, another three being added singly at random intervals afterwards. From January 1879 all the 5ft 0in and 5ft 6in locomotives were built with 17in by 22in cylinders, some of the latter built in 1882-83 having a modified type of frame with underhung springs for the coupled wheels.

Darlington Works built 25 bogie tanks of similar design between June 1877 and November 1880, 23 of which had 5ft 0in wheels (Plate 5.5), one had 5ft 6in diameter wheels and one had 5ft 8in diameter wheels. Gateshead Works provided the boilers for five of the locomotives built in 1878. From August 1879 the Darlington locomotives were built with 17in by 22in cylinders. (The 1886 designation of all these tank locomotives was Class BTP.) The bunkers of many of the 124 BTPs were later heightened by a wooden plank, and a small number had a single iron plate of similar depth. Subsequently many were given two coal rails, later plated up on the inside. Some retained wooden bogie brake blocks (seen in Fig 4.4b) after iron brake blocks had been substituted on the coupled wheels.

Goods Locomotives 1877-1882

An improved design of inside frames for Fletcher's 0-6-0 goods locomotives was tried on No 398 which was built at Gateshead Works in April 1877, the wheels being 5ft 0in diameter, the cylinders 17in by 24in and the wheelbase 16ft 6in as before. Instead of the previous deep, parallel-sided frames with large horizontal slots, it had solid frames, reduced in depth between the wheels, and all the springs were

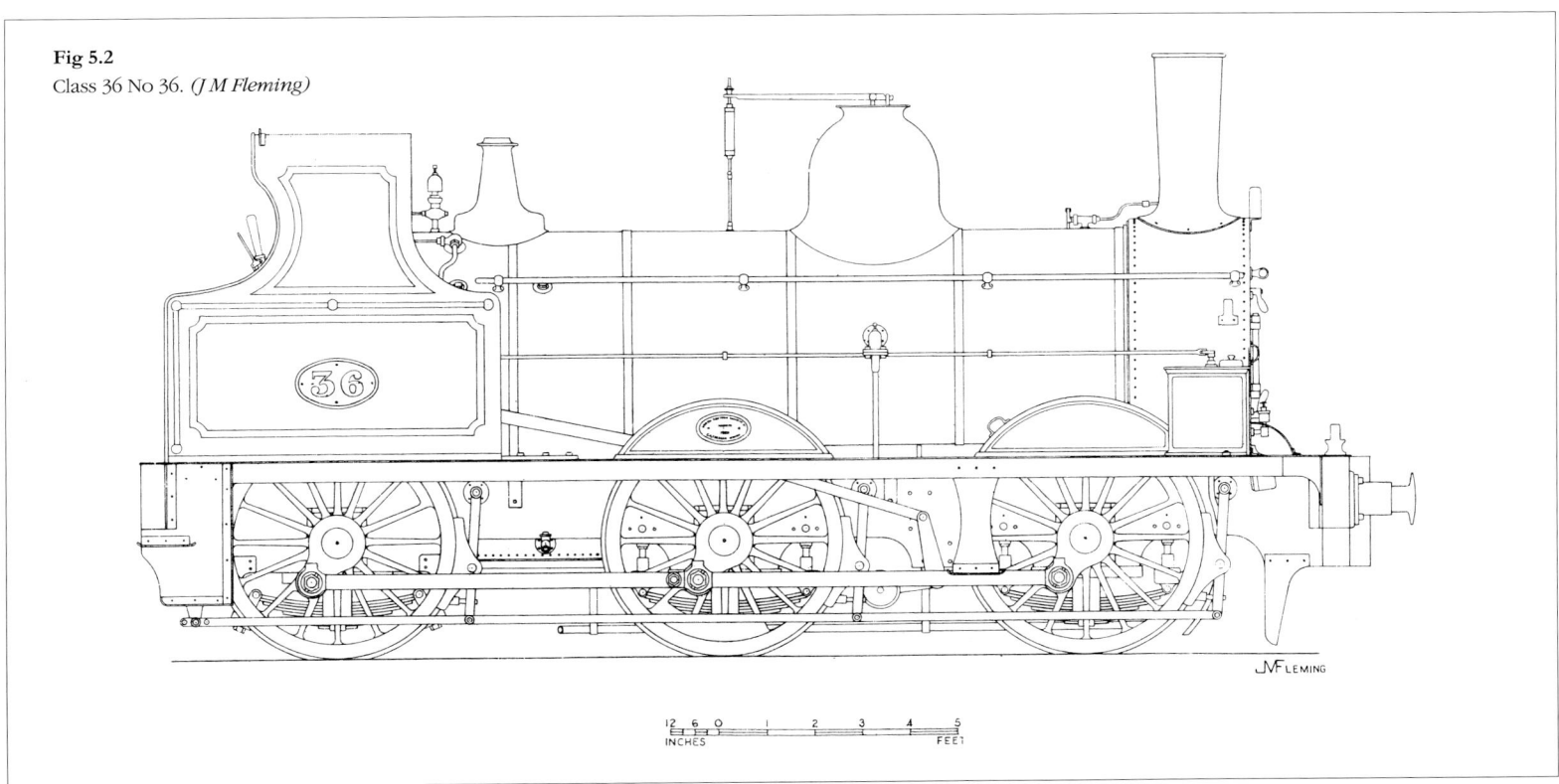

Fig 5.2
Class 36 No 36. *(J M Fleming)*

81

Plate 5.6
Class 398. No 292, seen in its original condition and livery, was built at Gateshead in May 1883, after the arrival of Alexander McDonnell but before he had introduced his new designs. Gateshead Works introduced a circular smokebox door, of the type illustrated, in 1883 and it was found on a few locomotives built or reboilered during the early part of that year before McDonnell introduced his archaic-looking double hinged doors.

underhung. The weighbar shaft was located below the motion, with the reversing rod passing through a slot in the right-hand driving wheel splasher. After some modifications the new design was adopted for future 0-6-0s and entirely superseded the earlier pattern of frames after April 1879 (Fig 5.1). The removal of the springs from their earlier position above the axleboxes allowed the boilers to be set lower on the frames. Brakes were fitted to the locomotives – both old and new – at about the same date to supplement those on the tenders.

The design was further improved by extending the frames by six inches at the rear to provide a better location for the bearings for the brake cross-shaft. Longer springs were also fitted, the hangers incorporating india-rubber cushions. Forty-eight of these longer 0-6-0s were built to Fletcher's design between December 1880 and May 1883, the last eleven being completed after Fletcher had retired (Plate 5.6). Another eight were turned out in 1883-84 with some alterations ordered by his successor. (The 1886 designation was 'Class 398'.)

A smaller version of these 0-6-0s was brought out in 1880. The wheel and cylinder dimensions were unchanged but the wheelbase was reduced to 15ft 8in (Fig 5.2). Only five locomotives were built, their numbers being: 36, 141, 229, 336 and 393 (The 1886 designation was Class 36.)

Darlington Works produced eight 0-6-0s built to a somewhat different design between June 1878 and March 1879. They were not all alike, although all had 5ft 0in diameter wheels, 17in by 24in cylinders, a 16ft 6in wheelbase, divided 8ft 0in + 8ft 6in and underhung springs. The first was No 12, which had frames that were similar to the latest

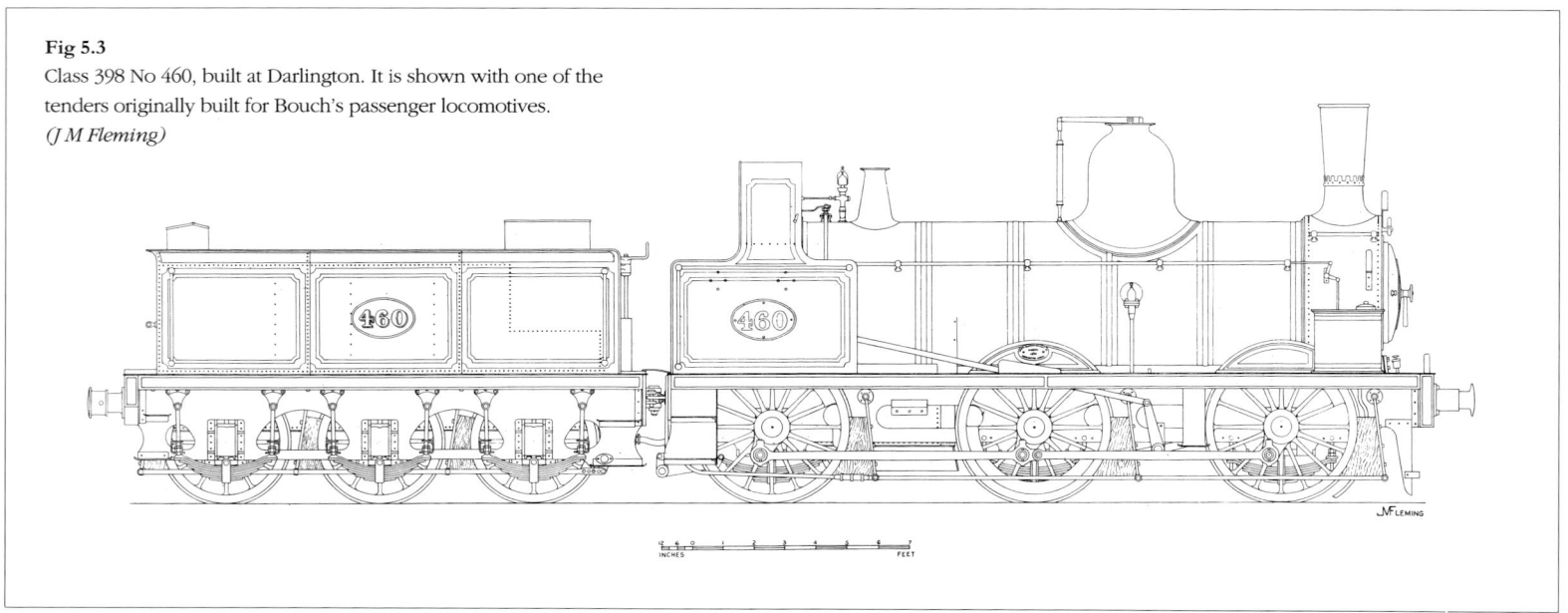

Fig 5.3
Class 398 No 460, built at Darlington. It is shown with one of the tenders originally built for Bouch's passenger locomotives.
(J M Fleming)

Plate 5.7
Class 124. No 609 is shown at Gateshead Works in its original condition. As on the BTPs, one can see the profile of the rear water tank in the curving line of rivets below the coal bunker.

Gateshead pattern with the lower edge cut back between the wheels. The last was No 589 about which no details are available. (Nos 12 and 589 were later included in Class 398.) The remaining six locomotives had slotted frames of the form that was typical of 0-6-0s built at Darlington between 1875 and 1882, their numbers being 358, 592/4 and 603/22/30. Despite their similarity to Class 398 they were separately designated Class 603.

Forty-one more 0-6-0 goods locomotives were built at Darlington between October 1879 and February 1883, 30 of which had 5ft 0in diameter wheels (Fig 5.3) and nine had 5ft 8in diameter wheels. All had 17in by 26in cylinders and a 16ft 6in wheelbase, divided 8ft + 8ft 6in. The weighbar shaft was below the motion but, unlike the Gateshead locomotives, the reversing rod passed through a guide on the outside of the driving wheel splasher. The splashers of the locomotives with 5ft 8in wheels were pierced by two pear-shaped openings. The last two locomotives, Nos 1087 and 1092, completed after Fletcher's retirement, were fitted with boilers manufactured at Gateshead Works. Incidentally, these were the last locomotives built at North Road until the first of the McDonnell 0-6-0s came out in September 1883. Several of the 2,400 gallon tenders built for Bouch's passenger locomotives were transferred to these 0-6-0s in exchange for their new 2,500 gallon Gateshead-pattern tenders. (These locomotives were later included in Class 398.)

Goods and Shunting Tank Locomotives

Between July 1881 and March 1882 a class of twelve 0-6-0Ts of very smart appearance was built at Darlington Works for local goods traffic

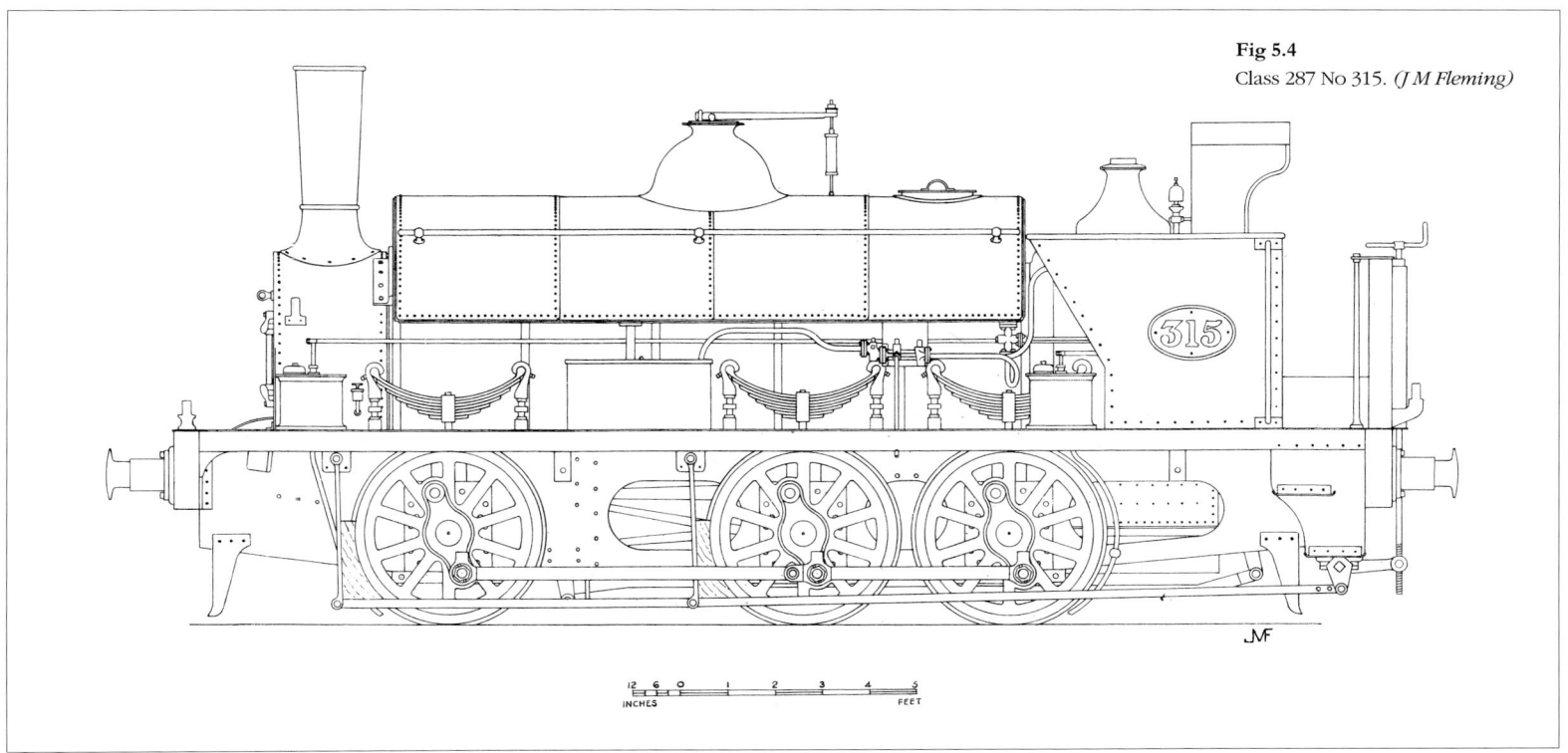

Fig 5.4
Class 287 No 315. *(J M Fleming)*

Plate 5.8
Class 44. One of the last Fletcher locomotives to be built, No 49 was completed in January 1883, after the arrival of Alexander McDonnell but unaffected by this event. Seen in its original condition, though in T W Worsdell livery, at Trafalgar Yard, Newcastle. It was scrapped in February 1925.

(Plate 5.7). The design owed much to the BTP 0-4-4Ts, having a similar boiler and a smaller version of the rear tank and bunker; a well tank was also added, increasing the total water capacity to 650 gallons. The wheels were 4ft 6in diameter originally, the cylinders 17in by 24in and the total wheelbase 15ft 8in; the wheelbase from leading to driving wheels was 7ft 8in which was the same as that of the 0-4-4Ts. The locomotives were built in two batches of six each, the second lot having larger leading sandboxes and underhung springs for the leading axles; all had underhung springs for the trailing axles. Some were fitted with Westinghouse brake equipment at quite an early date and were available for working passenger trains. Their water capacity was somewhat limited which led to No 211 being fitted with a saddle tank in 1897, later replaced with side tanks in 1912. All except Nos 609 and 610 were also fitted with side tanks which increased their water capacity to 1,280 gallons. (The 1886 designation was Class 124.)

Rebuilding of old locomotives with new frames, wheels and cylinders to become useful 0-6-0 saddle tanks continued throughout the 1870s, the most numerous being the short-coupled locomotives that were later designated Classes 287 and 48 (Fig 5.4).

The last NER saddle tank shunting locomotives were seven 0-6-0STs built at Gateshead Works between December 1881 and April 1883, the first one being No 44. They bore some resemblance to the locomotives built by R & W Hawthorn & Co in 1875-76 but had 4ft 6in diameter wheels and 15in by 24in cylinders, the wheelbase being 6ft 6in + 7ft 1in. (One engine, No 98, had 4ft 0in diameter wheels). When new they were turned out in passenger livery with all the panelling and lining-out of the period (Plate 5.8). (The 1886 designation was Class 44.)

Chapter 6

Locomotive Development 1883 – 1885
Alexander McDonnell, Locomotive Superintendent

Alexander McDonnell came to the NER from the Great Southern & Western Railway of Ireland (GS&WR) where he had reorganised the Locomotive Department with considerable benefit to that company. His tenure of office as Locomotive Superintendent of the NER lasted less than two years during which time he brought about many changes, including the enlargement and modernisation of the workshops. He also introduced new liveries for the locomotives.

McDonnell took up his appointment with the NER on 1 November 1882 but the first locomotives built to his design did not appear until September 1883. Meantime both Gateshead and Darlington had continued to build new locomotives to existing designs, even perpetuating the Fletcher liveries at least until as late as May 1883, and possibly for a few months longer.

Five engines of a final batch of eight modified Fletcher 0-6-0s with closed domes, Ramsbottom safety valves and McDonnell's cast-iron numberplates, were completed at Gateshead in October/November 1883.[1] The much-needed reorganisation of Gateshead Works interrupted progress with the last three engines; the components which had already been made were sent to York Works to be erected; they were completed in September 1884 with additional modifications.

A final order for six 0-4-4Ts was also delayed. Four were completed at Gateshead Works between June and October 1883, the fourth one having McDonnell's boiler mountings and numberplates. The components for the last two were sent to York Works where they were completed in December 1884 with McDonnell cabs and extended tanks, although having pre-1883 'York-pattern' boiler mountings.

The two new classes of locomotives introduced by McDonnell in 1883-84 departed from traditional NER practice in several respects and these made them unpopular with the drivers. Most of the engines were not actually turned out until after McDonnell's sudden departure from the NER on 19 September 1884, construction continuing until well into 1885.

---

[1] A number of other Fletcher locomotives had their number plates replaced by cast-iron ones of the McDonnell pattern.

Plate 6.1
Class 59. No 235, one of the first batch, built at Darlington in January 1884.

Goods Locomotives 1883-1885

McDonnell's NER locomotives resembled those which he had previously designed for the GS&WR. Darlington Works built 32 of his Class 59 goods engines between September 1883 and September 1885, another 12 being supplied by R Stephenson & Co in 1884-85. They had 17in by 26in cylinders, 5ft 0in wheels (two had 5ft 6in wheels) and a 16ft 0in wheelbase, divided 7ft 9in + 8ft 3in. The running plates were only four feet above rail level, those of the Darlington engines having raised 'humps' to provide clearance for the oil cups on the coupled wheels (Plate 6.1); the wheels of the Stephenson engines had the crank throw reduced to only ten inches to avoid the need for 'humps'.

Other innovations included smokeboxes with sloping fronts and pairs of semi-circular doors (Plate 6.2), chimneys decreasing in diameter towards the top, closed domes and Ramsbottom safety valves. The cabs had wooden roofs, large windows in hinged wooden frames and cast-iron numberplates with polished numerals and borders. The engines were driven from the left-hand side and they had screw reversing gear instead of Fletcher's convenient lever-and-screw arrangement; also they lacked exhaust cocks.

All but the last eight Darlington engines had 'NORTH EASTERN RAILWAY' engraved in the brass beadings of the driving wheel splashers; the narrow beadings of the first eight required wider insertions to accommodate the lettering. Most locomotives had leading sandboxes of the same height as the splashers with which they were combined; the last sixteen Darlington engines could be distinguished by their taller sandboxes. (Plate 6.3) Tenders with narrow tanks and with springs above the running plates were built for the Darlington engines; the Stephenson engines had wide tanks and springs below the running plates.

Passenger Locomotives 1884-1885

Between February and September 1884 Gateshead Works completed the first eight of McDonnell's Class 38 4-4-0 passenger locomotives (Plate 6.4 and Fig 6.1). They had steel frames, swing-link bogies – a type of bogie not much used subsequently by the NER – and volute springs for all axles. The cylinder dimensions were 17in by 24in, the coupled wheels 6ft 7in diameter, bogie wheels 3ft 0in diameter and wheelbase 5ft 3in + 6ft 11½in + 8ft 4in, the total wheelbase being 20ft 6½in. The boilers were 10ft 3in long by 4ft 3in diameter; the firebox shells were 5ft 6in long. The running plates were raised to clear the coupling rods.

A second order for eight more 4-4-0s was completed between October and December 1884 but it was soon found that the engines lacked sufficient power to meet NER requirements and a third order, on which work had begun, was cancelled.

An additional twelve 4-4-0s were delivered by R & W Hawthorn & Co between October 1884 and April 1885 after McDonnell's departure. They differed from the Gateshead products in several respects, these including a coupled wheelbase four inches longer, laminated springs for the crank axle, helical springs for the rear axle and a gentler curvature of the running plates at each end of the raised sections. (Plate 6.5)

The tenders of the Hawthorn 4-4-0s were similar to those built for the Stephenson 0-6-0s. The Gateshead-built 4-4-0s had tenders with shorter footplates at the front and shorter handrails supported by only one pillar instead of two. Several of the Gateshead engines exchanged tenders with Fletcher engines in order to reduce the overall wheelbase. No 186 when newly built was given the latest type of 2,500 gallon tender but most engines received older 1,862 gallon tenders; No 281 had a 1,500 gallon tender of the type also seen with 2-4-0 No 64.

Plate 6.2
Class 59. No 78, built at Darlington in December 1884, showing the sloping front to the smokebox and sandbox and McDonnell's use of a pair of smokebox doors.

Plate 6.3
Class 59. No 78, illustrating two differences in the appearance of the later Darlington engines. A larger sandbox has been provided, which now projects above the leading splasher, while a broader brass beading has been employed on the second splasher – so there is no longer a bulge to accommodate the company name.

Plate 6.4
Class 38. No 186, built at Gateshead and photographed when brand new in McDonnell's brown livery. Its boiler pressure of 140psi was typical for the period.

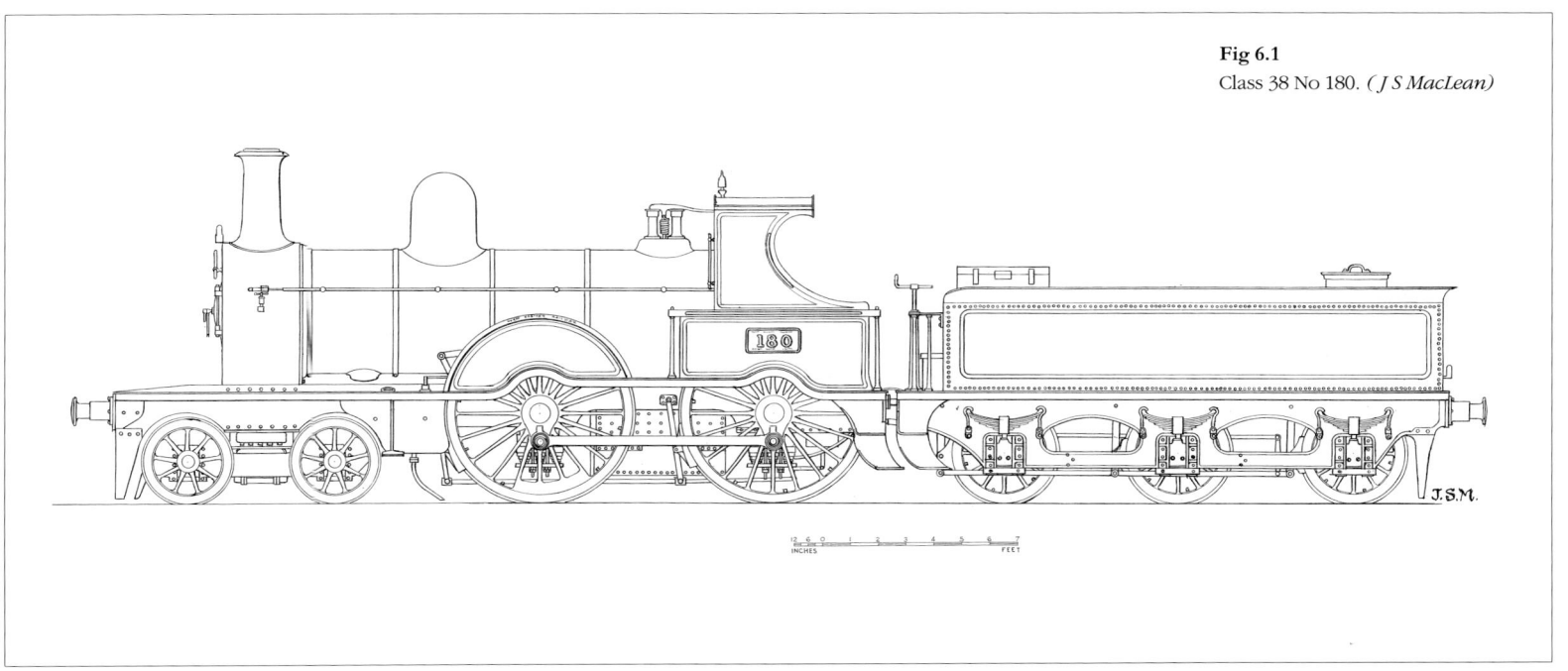

Fig 6.1
Class 38 No 180. *(J S MacLean)*

Plate 6.5
Class 38. No 1496, one of the Hawthorn-built locomotives. The most visible difference from those built at Gateshead is the gentle upsweep of the running plate.

McDonnell began a programme of replacing worn-out boilers of existing engines with new ones having closed domes[2] and Ramsbottom safety valves; some serviceable boilers were also altered to conform to his standards. The driving position of some of Fletcher's locomotives was changed from right to left-hand, the exhaust cocks were removed, new chimneys, cabs, splasher beadings and number plates were provided and coal rails were added to the tenders. However, comparatively few engines were subjected to all these cosmetic changes, thereby adding further variety to the NER scene.

Plate 6.6
Class 38. No 180, built at Gateshead, has received a Worsdell chimney and is painted in T W Worsdell livery. The McDonnell tender had 2,500 gallons water capacity and 3ft 8 ½in wheels.

Several of Fletcher's 901 class 2-4-0s had their leading springs moved to an inaccessible position out-of-sight below the running plates, the alteration being accompanied by modified frames and the removal of the leading footsteps.

Alexander McDonnell's Liveries

McDonnell adopted a dark brown livery, probably in September 1883 when his first 0-6-0 was completed. The colour was described in *The Locomotive Magazine,* April 1910, page 72, as ". . . similar to that in vogue on the North Staffordshire Railway until a few years ago, the lining being red and white." (The North Staffordshire Railway had adopted 'Victoria' brown – a very dark shade of red-brown – in 1882). J. Brown remembered seeing 4-4-0 No 186 in the dark brown livery when newly built in April 1884; he also saw some of Fletcher's

---
2   In contrast to having safety valves on the dome.

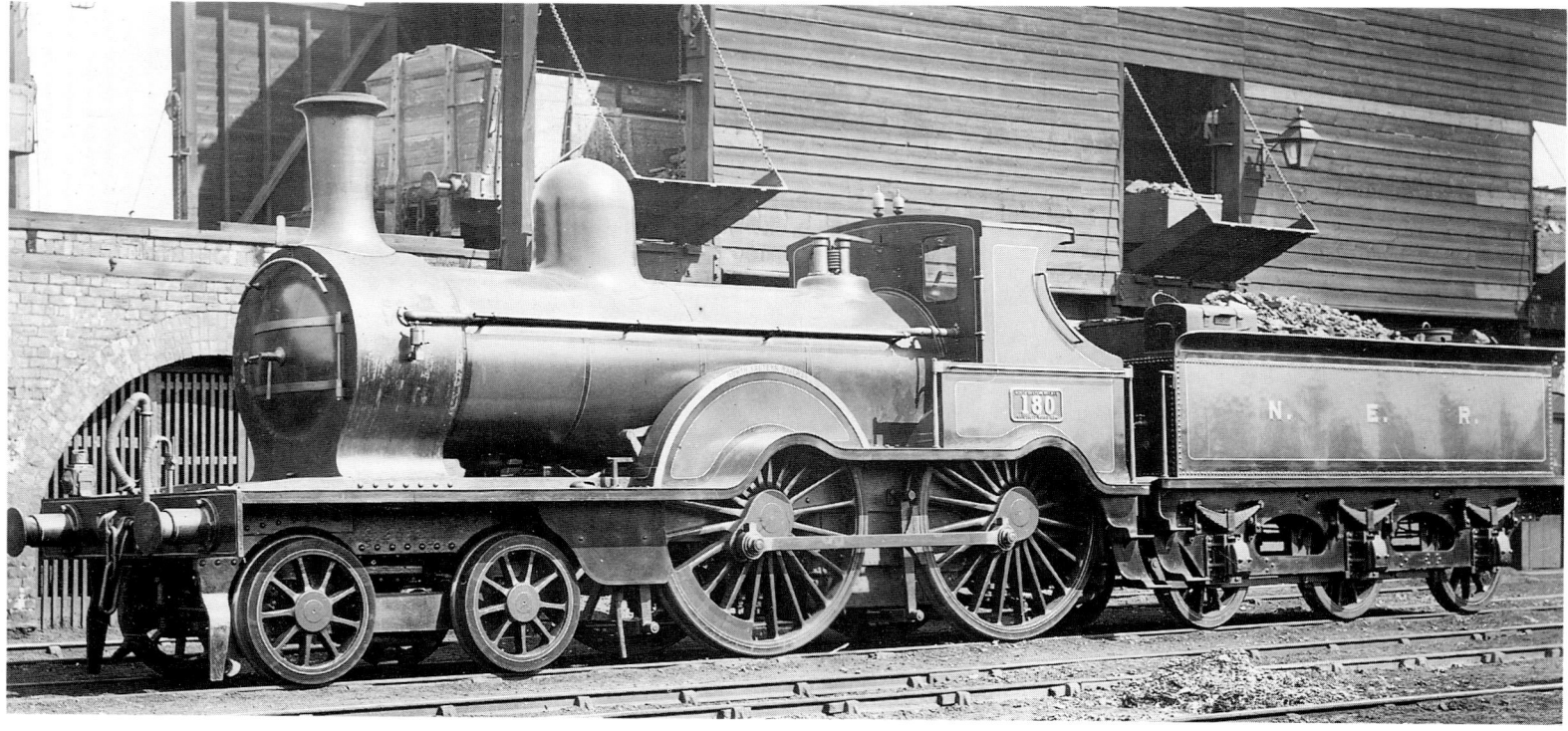

2-4-0s – Nos 362 and 925 – painted brown after their alteration by McDonnell during 1883-84 (he described this to the author *c*1935). In *Moore's Monthly Magazine* (forerunner of the *Locomotive Magazine*), December 1896, page 140, the colour was described as dark chocolate.

The description in *The Locomotive Magazine*, quoted above, also stated that McDonnell later ". . . adopted a dark blue *(sic)* livery (later corrected to olive green) with black bands and fine red and white lines, the inside of the cab being grained in light oak and the underframe claret colour."

It seems that McDonnell's green livery was similar to that used by the GS&WR which was a medium olive green with black panel bands bordered on each edge with a light green line, and with a fine vermilion line on the inner side; the boiler cleading belts were black with a light

The 'Tennant Interregnum': September 1884 to August 1885

Between McDonnell's resignation and the appointment of a successor the NER Locomotive Department was supervised by a committee chaired by the company's General Manager, Henry Tennant, who proposed that an enlarged version of Fletcher's Class 901 2-4-0s should be built to meet an urgent need for more powerful locomotives for the heavier passenger trains. The result was the construction of the Class 1463 2-4-0s popularly known as the *Tennants*, twenty of which were built between May and September 1885, Nos 1463-72 at Darlington and Nos 1473-79 and 1504-06 at Gateshead (Plate 6.8 and

Plate 6.7
2-2-2 No 1709, built by Robert Stephenson & Co in December 1853 as No 159 and renumbered 1709 on 1 February 1885. Seen with a McDonnell chimney, recently ex-works in his green livery. The sheen picked up in this photograph comes from tallow being rubbed on to the paintwork.

green line and fine red line on each side. The frames were dark red brown with a narrow black border, having a fine green line on its inner edge and a vermilion line inside it; the buffer beams were red with black borders, black buffer sockets and yellow numerals shaded with blue. The numberplates had polished cast-iron letters and figures on a black ground. The cabs were painted a light straw colour inside.

The above description fits the livery seen in several photographs of NER passenger engines *c*1884-85, eg 0-4-4T No 954 and 2-2-2 No 1709 (formerly No 159). After mentioning the McDonnell chimney and closed splashers fitted to No 1709 in 1884 (Plate 6.7) E L Ahrons referred to "McDonnell's peculiar dark green livery".[3] Goods engines appear to have had a less elaborate lining out.

Fig 6.2). Among those responsible for the design were Wilson Worsdell, Assistant Locomotive Superintendent at Gateshead since 1883, Mr. Hepper, previously Edward Fletcher's Chief Draughtsman for new work, and George Graham (later appointed Assistant Locomotive Superintendent).

The new engines had 4ft 7¼in diameter leading wheels, 7ft 1¼ diameter coupled wheels, an 8ft 0in + 8ft 8in wheelbase and 17in by 24in cylinders. The upward sweep of the outside frames, which avoided the need for separate coupling rod splashers, may have been suggested by Tennant himself. The engines were driven from the right-hand side and had combined lever-and-screw reversing gear; exhaust cocks were

---
3 Letter to J S MacLean 13 December 1913

also re-introduced. The tenders designed for the *Tennants,* holding 2,651 gallons, were the prototype for subsequent NER tenders, the size and capacity being increased from time to time. Their 3ft 9¼in wheels also remained standard for the majority of later tenders.

The Saxony green colour of Fletcher's later years was re-introduced but without Brunswick green borders. The cleading belts were black, with a ⅛in white line on each edge, and there were black borders, with ⅛in white lines around the cabs, splashers and the sides and rear of the tenders; in addition, the tenders had rectangular panels formed by 2in black bands between ⅛in white lines. Gold numerals, shaded red, were used for the engine numbers on the rear splashers, with similar letters for the initials N E R on the tenders. Maker's plates of polished brass were carried on the leading splashers, but the *Tennants* never had numberplates until after the Railway Grouping in 1923. The outside frames of engines and tenders were dark claret colour, the Darlington Works using a slightly darker shade than Gateshead.

Plate 6.8

Tennant The first of the class, No 1463, built at Darlington in 1885 and seen in its original condition. It was preserved by the LNER and is now on display at North Road Station in Darlington.

Plate 6.9

Class 8, No 461. In addition to the *Tennants*, the committee oversaw the building of eight 0-6-0Ts at Gateshead for shunting duties. The first, No 8, appeared in August 1885; No 461 in October. They used some parts which had originally been ordered for further Class 38 4-4-0s. The fronts of the side tanks had a distinctive semi-circular profile. *(J F Mallon collection)*

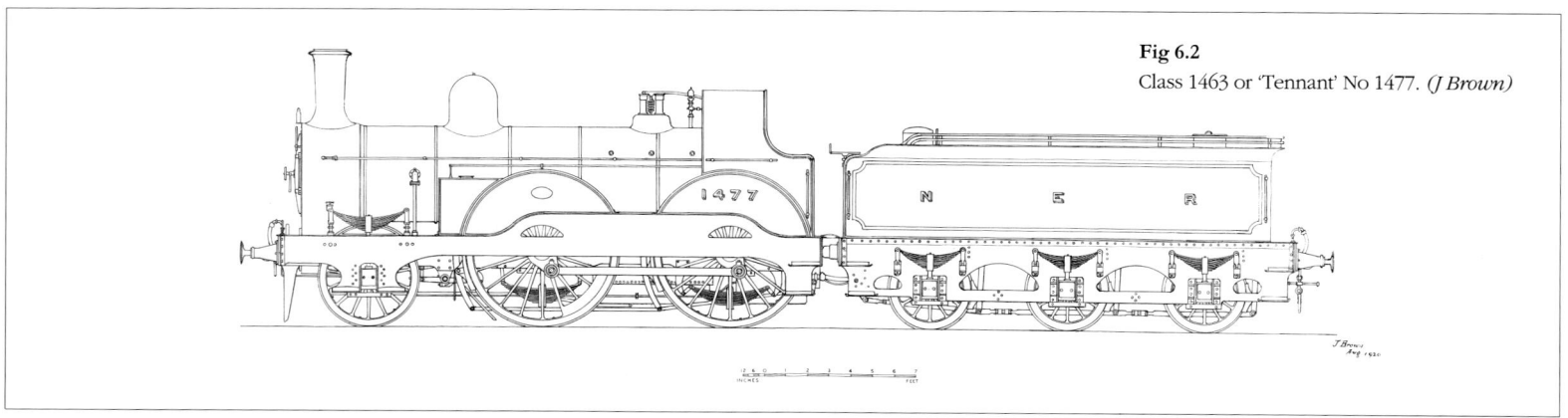

Fig 6.2
Class 1463 or 'Tennant' No 1477. *(J Brown)*

Chapter 7

Locomotive Progress 1886 – 1890
Thomas William Worsdell, Locomotive Superintendent

The interregnum in the Locomotive Department came to an end in September 1885 when Thomas William Worsdell took up his appointment as Locomotive Superintendent (having been appointed on 17 April with effect from 1 September). He had served the Great Eastern Railway (GER) in a similar capacity since 1882 and, before that, had been Works Manager, under Francis Webb, at the Crewe Works of the London & North Western Railway (LNWR) from 1871.

Modernisation and Standardisation

NER locomotive practice underwent a complete transformation during T W Worsdell's regime. He further extended the use of modern machine tools in the workshops, thereby achieving greater accuracy and interchangeability of parts. Extensive use was made of steel, not only for frames and boilers but also as castings for wheel centres, motion plates, firebox roof stays and various frame brackets and fittings for which forged iron had previously been employed. Annealing furnaces and hydraulic presses were installed for flanging the boiler plates.

Productivity was also improved by standardisation of many parts of the locomotives, including the introduction of a standard range of boilers to fit both new and many existing classes. The majority of the 'main line' locomotives were built as two-cylinder compounds on the Worsdell and von Borries system and Joy's radial valve gear was adopted for all but one new class of locomotive.

T W Worsdell's locomotives were notable for their extremely neat, 'Golden Age' appearance and for the much improved protection given to the enginemen by the provision of more commodious cabs. Decorative brasswork was limited to a handsome cover for the Ramsbottom safety-valves, a collar around the front of the boiler, the numberplates and, in the case of passenger tender locomotives, the splasher beadings.

An alphabetical system of classification was adopted for new locomotives, with the addition of a numerical suffix to denote a variation from, or a development of, a basic design; where both compound and simple versions of a design existed, the compound was deemed to be the principal class. Pre-1886 locomotives were allocated numerical classifications related to a typical example of each type – eg Classes 38, 901 and 1463.

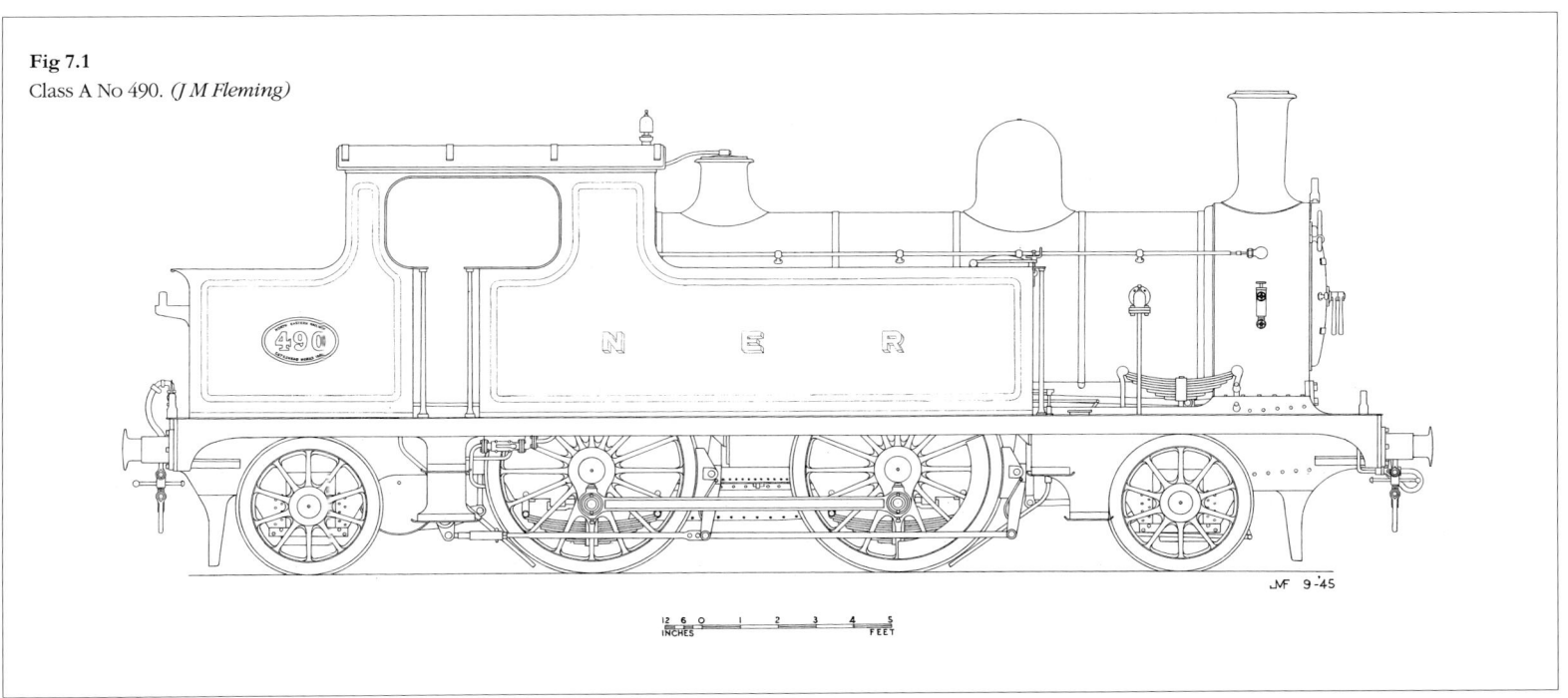

Fig 7.1
Class A No 490. *(J M Fleming)*

New Standard Locomotives

T W Worsdell's first locomotives for the NER were the Class A 2-4-2Ts which were introduced in March 1886 for local passenger trains (Plate 7.1 and Fig 7.1). They had 5ft 7¼in diameter coupled wheels, 18in by 24in cylinders and a slightly larger boiler than Worsdell's 2-4-2Ts for the GER, but in many respects the two designs were exactly alike.[1]

The crank axles for the first ten locomotives had oval webs, with hoops shrunk on for extra strength, but Worsdell afterwards introduced an improved form of axle with circular webs which could be finished on a lathe; this design remained the standard pattern for all subsequent NER inside-cylinder locomotives. Radial axleboxes were provided for the leading and trailing axles of the 2-4-2Ts with laminated springs to control the lateral movement (Fig 7.2). The pump for the Westinghouse brake was housed inside the cab, at the rear of the left-hand side tank. It is of interest to note that the 5ft 6in diameter wheels of Fletcher's 2-4-0s of 1870-75 were modified for re-use on the 2-4-2Ts built in 1887. There were eventually 60 locomotives in Class A.

Five more new types of locomotive were introduced in 1886, two of which were built as two-cylinder compounds.

---

1. The first 20 locomotives of Class A had 17in by 24in cylinders, later enlarged to 18in. The boiler pressure was 160psi.

Plate 7.1
Class A. No 674 seen in front of Esk Terrace, Whitby, in T W Worsdell livery with the square stops to the initials introduced from early 1887.

The need for more powerful goods locomotives was met by the Class C1 0-6-0s and the Class B1 0-6-2Ts (Plate 7.2), ten of each class being built in 1886. As far as possible the design and dimensions of the two classes were identical, both having 5ft 1¼in diameter coupled wheels and 18in by 24in cylinders.

Compounds

There followed two experimental two-cylinder compounds, comprising one Class C 0-6-0 (Plate 7.3 & Fig 7.3) – a compound version of Class C1 – and one Class D 2-4-0, with 6ft 8¼ diameter coupled wheels, for express passenger traffic (Plate 7.4 and Fig 7.4). Both locomotives had an 18in diameter high-pressure cylinder on the left-hand side and a 26in diameter low-pressure cylinder on the right, the stroke being 24in in each case. A differential adjustment of the valve gear quadrants gave a later cut-off in the low-pressure cylinder than in the high-pressure cylinder in order to equalise their work

Plate 7.2
Class B1. No 428 apparently photographed in grey but with panels and lining corresponding to its original livery.

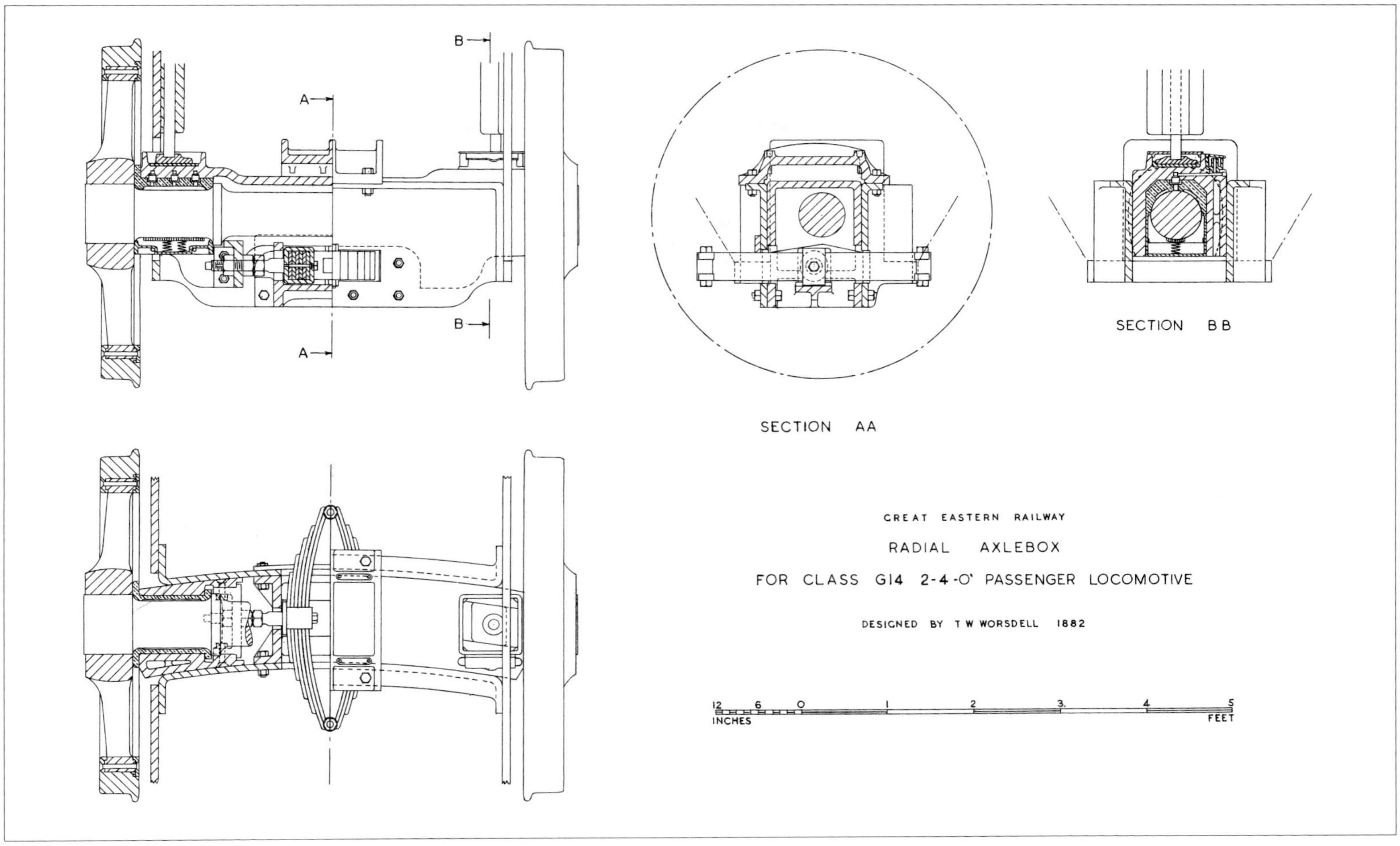

Fig 7.2
T W Worsdell's radial axlebox, designed for Great Eastern Railway Class G14 2-4-0 and used on NER Class A. *(J M Fleming)*

Plate 7.3
Class C. No 666 in Gateshead livery, with the upper and lower panels of the cab side lined separately. Locomotives painted at Darlington had the entire cab side lined as a single panel with additional lining around each window.

Plate 7.4
Class D. No 1324 in its original form as a 2-4-0 compound. It is shown with the very early change to the tender livery, in which the company's initials were replaced by 'NORTH' and 'EASTERN' flanking the T W Worsdell carriage-style armorial device set within a garter. It was rebuilt as a 4-4-0 simple and reclassified F1 in October 1896.

Fig 7.3
Class C No 16. *(J M Fleming)*

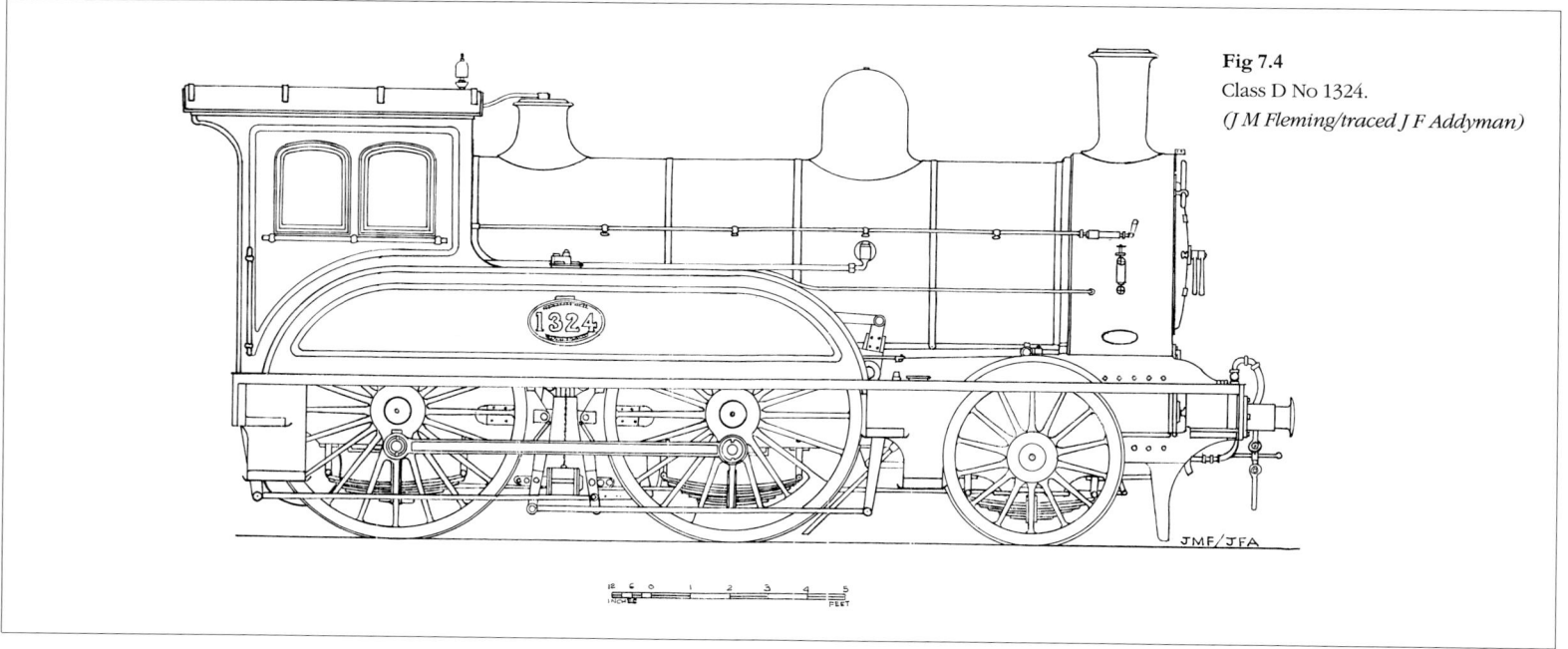

Fig 7.4
Class D No 1324.
(J M Fleming/traced J F Addyman)

Plate 7.5
Class E. No 275 at Redheugh Quay, Gateshead.

as nearly as possible. The occasional problem of starting a two-cylinder compound when the high-pressure piston was was at the end of its stroke was overcome by means of Worsdell's 'Automatic Starting and Intercepting Valve' (Fig 7.5) which admitted live steam to the low-pressure cylinder and enabled the compounds to be started as easily as a simple locomotive.

The first phase of Worsdell's standardisation programme was completed in November 1886 with the introduction of the Class E 0-6-0T shunting locomotives (Plate 7.5 and Fig 7.6). This was T W Worsdell's only class of NER locomotives to be fitted with Stephenson link motion.

External Characteristics

All these locomotives had a very 'modern' appearance in comparison with most of their contemporaries, and something of this 'Worsdell Style' could be seen in all subsequent NER locomotive designs. The combined splashers over the driving and coupled wheels of the 2-4-0 – and later 4-4-0 – designs enclosed the Westinghouse brake pump on the right-hand side of the locomotives, access to the pump being through an opening in the side of the splasher which was covered by a hinged flap carrying the number plate. The splashers were extended within the cab as flat-topped iron boxes, the box on the right-hand side serving to support the reversing screw

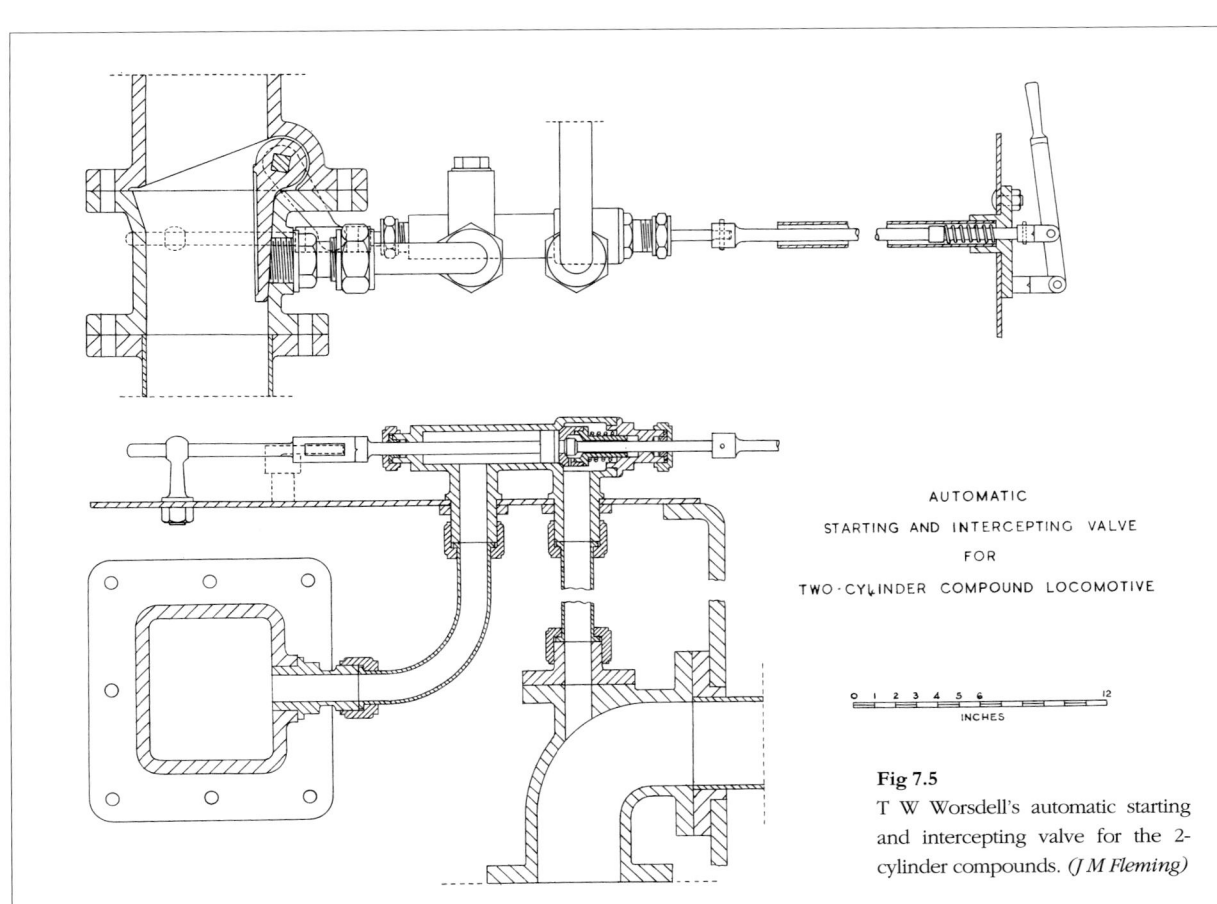

Fig 7.5
T W Worsdell's automatic starting and intercepting valve for the 2-cylinder compounds. (*J M Fleming*)

whilst still leaving enough room for a wooden seat for the driver; the seat on the fireman's side occupied the full length of the cab.

Livery

During T W Worsdell's regime all classes of locomotives were given the same livery, ie Saxony green with claret coloured borders, the two colours normally being separated by a black panel band 2in wide, having a white line – nominally $\frac{1}{8}$in wide – on the inner edge and a vermilion line on the outer. The width of the claret borders was varied to suit the contours of the locomotive, being reduced where space was limited, as for example at the rear of the cab opening on the tank locomotives (Fig 7.1) and around the splashers and sandboxes of the goods and shunting locomotives. Where space was even more restricted, as around the side windows of the Class C and C1 cabs, the colours were separated by a single white line. The beadings around the side windows of the Class D cab were claret colour, with a single white line. Numberplates were black, with raised, polished brass numerals and lettering. Each compound had an additional brass plate lettered thus:-

No. . . .
PATENT COMPOUND LOCOMOTIVE
WORSDELL & v BORRIES
SYSTEM

Cab interiors above the seats were painted and grained to simulate wood, the lower part being black. Firebox backplates were also black, with a broad, polished brass beading. The outside surfaces of cab roofs were black. Boiler cleading belts were black, edged with a ¼in white line. Usually only the rear edge of the front belt was lined, and there was no lining at the junction between the boiler cleading and the front of the cab. The upper surface of the splashers was green, with a claret border and a white line; in the case of the passenger locomotives, the inner border and line both ended where the splashers joined the boiler lagging. The underside of boilers was painted black for a width of about three feet, and there was usually a narrow black strip along the bottom edge of splashers, cabs, tanks and tenders where the paintwork was liable to be damaged by enginemen's boots.

At first, the company's initials 'N E R' were shown on the tenders and side tanks in 6in gold letters, blocked red, shaded black and brown. From early in 1887, a square 'stop' was added after each letter. The locomotive number was shown in 4in gold figures at the back of the tenders, on the curved coping at the top.

The angle irons or valances below the running plate were claret colour, with a 1in black border along the lower edge and a ¼in vermilion line. Footsteps below the running plate were also claret colour, with black edges and vermilion lines; the treads were black.

Buffer beams were vermilion, with a 1in black edge and a ¼in white line; a rectangular black panel edged with a white line surrounded the draw hook which, with its links, was also black. The locomotive number was shown to the right-hand side of the draw hook in 4in gold block figures, shaded black or dark blue, 'No' being on the left-hand side in similar characters. The buffer stocks were vermilion, with a 1in black edge and a white line. Buffer heads and shanks were black.

Wheels were green, with black tyres and a white line. Axle ends were usually black, but sometimes green. The locomotive frames were black on their outer side and red on the inner, together with the motion plate, axles, connecting rods and those parts of the valve gear that were not left unpainted. The external parts of the reversing rods of the 0-6-0s were also – possibly – red.

Smokeboxes and chimneys were black, with the hinges, sealing ring and handles of the smokebox door unpainted and burnished. Cab, boiler and tender handrails were usually painted, although sometimes polished.

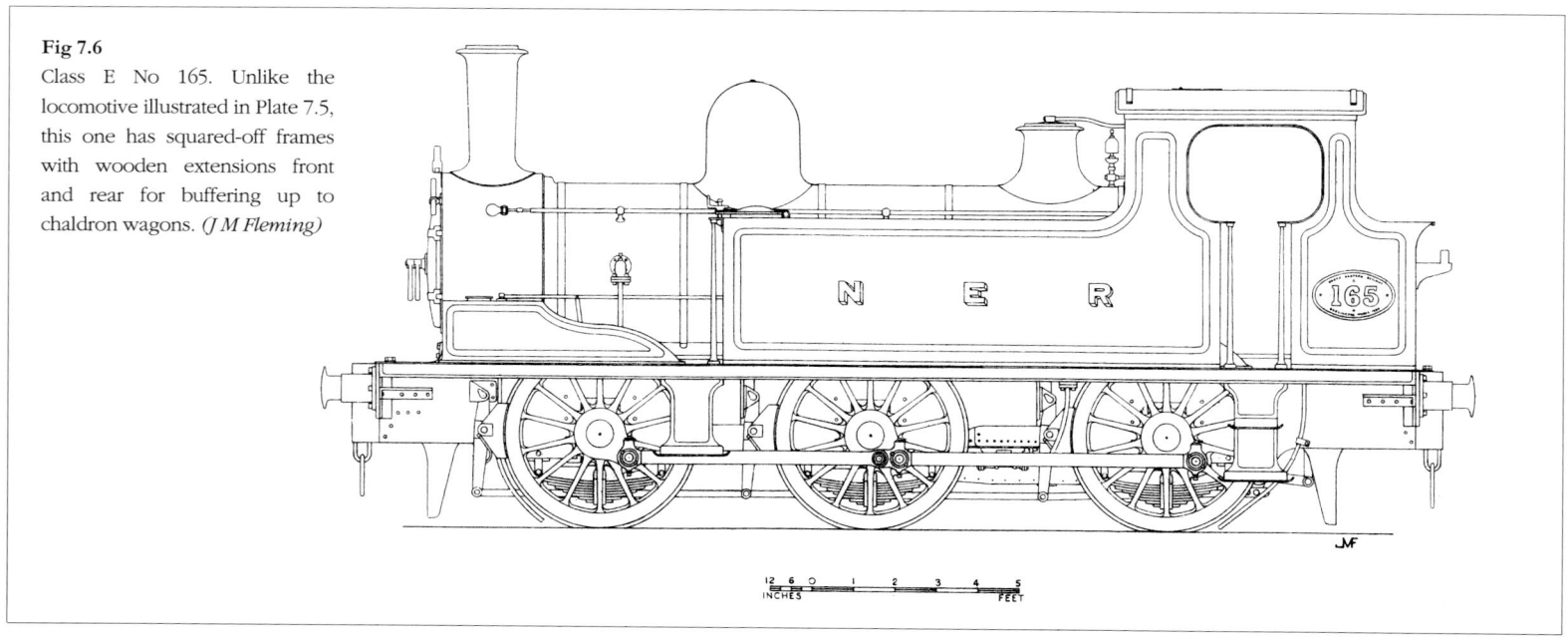

Fig 7.6

Class E No 165. Unlike the locomotive illustrated in Plate 7.5, this one has squared-off frames with wooden extensions front and rear for buffering up to chaldron wagons. *(J M Fleming)*

Plate 7.6
Class F. No 117.

Tender underframes were claret colour, with a 1in black border and ¼in vermilion line along the lower edge and around the footsteps, hornblocks and the 'half-moon' openings in the frames. Hornblocks, axleboxes and footsteps were black. The front and top surfaces of the tender, including the coal space and coal rails were black.

After the Class D 2-4-0 – No 1324 – had been at work for only a few months it was taken into Gateshead Works and given a 'special' livery, prior to being shown at the Royal Jubilee Exhibition at Newcastle, from May to September 1887. The colours were unchanged, but the initial letters 'N E R' on the tender were replaced by the company's heraldic device, flanked by 'NORTH' and 'EASTERN' in 4in gold letters, blocked red shading to brown. The paint was removed from the whole of the smokebox door and the top of the chimney, after which they were burnished, as were the buffer heads, wheel rims, valve gear and all handrails and brasswork. The interior of the cab was painted white above the level of the splasher boxes.

The style of lettering and heraldry first seen on the exhibition locomotive was adopted in 1888 for the tenders of all post-1885 passenger locomotives. Less distinguished locomotives continued to have nothing more than 'N. E. R.'

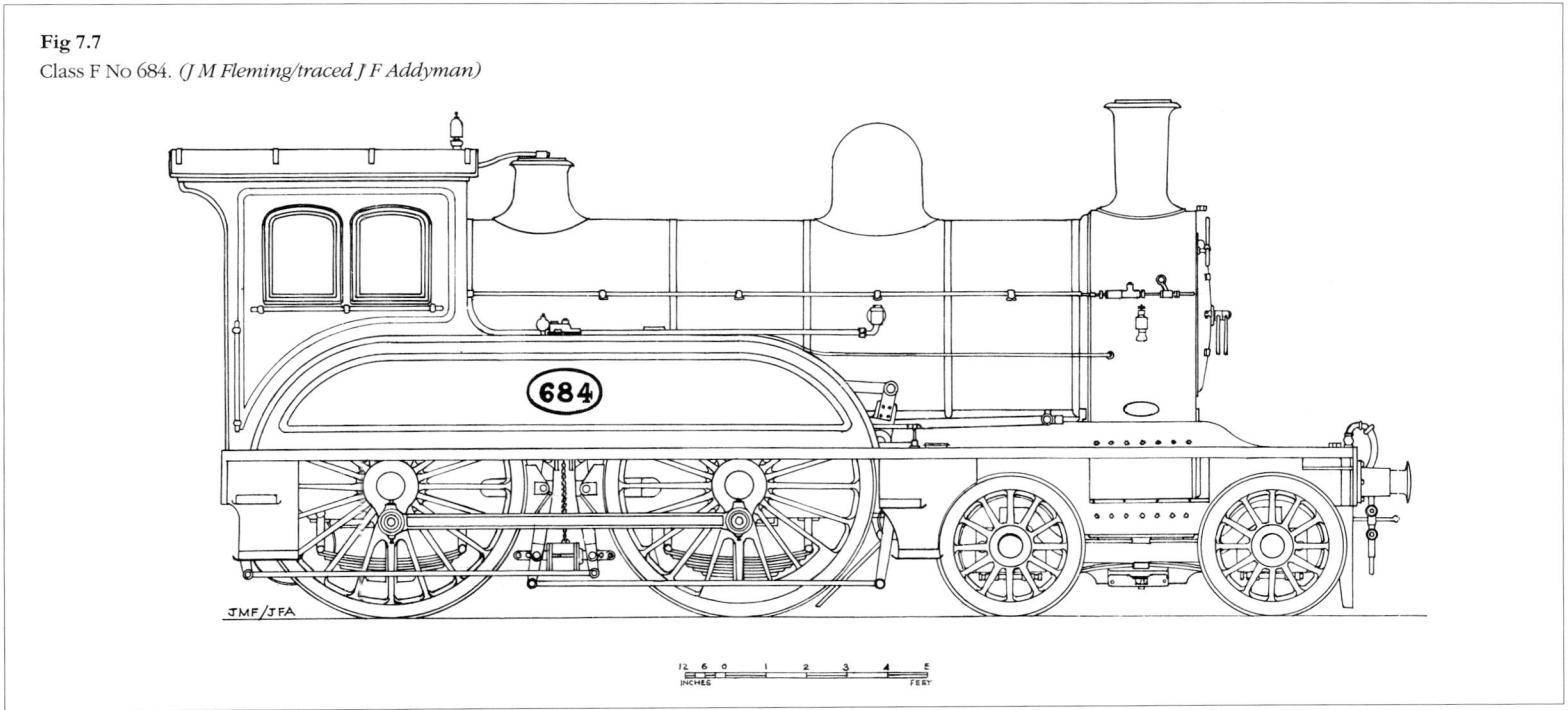

Fig 7.7
Class F No 684. *(J M Fleming/traced J F Addyman)*

Further Passenger Locomotives: Classes F, G and B

Further additions were made to Classes A, C and E in 1887 and succeeding years, whilst for express passenger traffic there was a new 4-4-0 design, identical with Class D except for the incorporation of a bogie in place of the single pair of leading wheels (Fig 7.7). The centre pin of the bogie entered a block which moved between curved guides, the lateral movement being controlled by double elliptical springs (Fig 7.8). The centre of curvature of the guides coincided with the centre of the driving axle, so that there was no rotation about the bogie pin. Buffers at the front of the bogie frame limited its maximum movement by coming into contact with the main frames of the locomotive. A slight bend in the main frames, ahead of the cylinders, gradually reduced their distance apart from 4ft to 3ft 9½in at the buffer beam. Ten of these 4-4-0s were built in 1887 as compounds (Class F) (Plate 7.6) and ten as simples with two 18in by 24in cylinders (Class F1).

While Gateshead Works was turning out the 4-4-0s, Darlington was starting to build the Class G1 2-4-0s for country branch line services which required something less powerful, and cheaper, than Classes F and F1. Class G1 had 6ft 1¼in coupled wheels and 17in by 24in cylinders, with a boiler and other parts common to the Class A 2-4-2Ts (Plates 7.7 & 7.8 and Fig 7.9).

Early in 1888 one Class B compound 0-6-2T was built, together with one more Class B1 simple, for comparison. The main section of their frames, which were 4ft apart, terminated three feet behind the trailing coupled axle, the rearward extension being made of separate plates spaced only 3ft 6in apart in order to allow greater lateral movement of the radial axle. The two parts of the frames were united by a substantial steel casting. (The original ten Class B1 0-6-2Ts had frames in one piece, spaced 4ft apart from end to end.) No further additions were made to Class B1, but 50 more of the compounds were built (Plate 7.9 and Fig 7.10).

One more Class D 2-4-0 compound – No 340 – was built in December 1888. It was the first locomotive to have piston valves of W M Smith's patent design, there being one 7in diameter valve above the high pressure cylinder and two 5½in diameter valves, side by side,

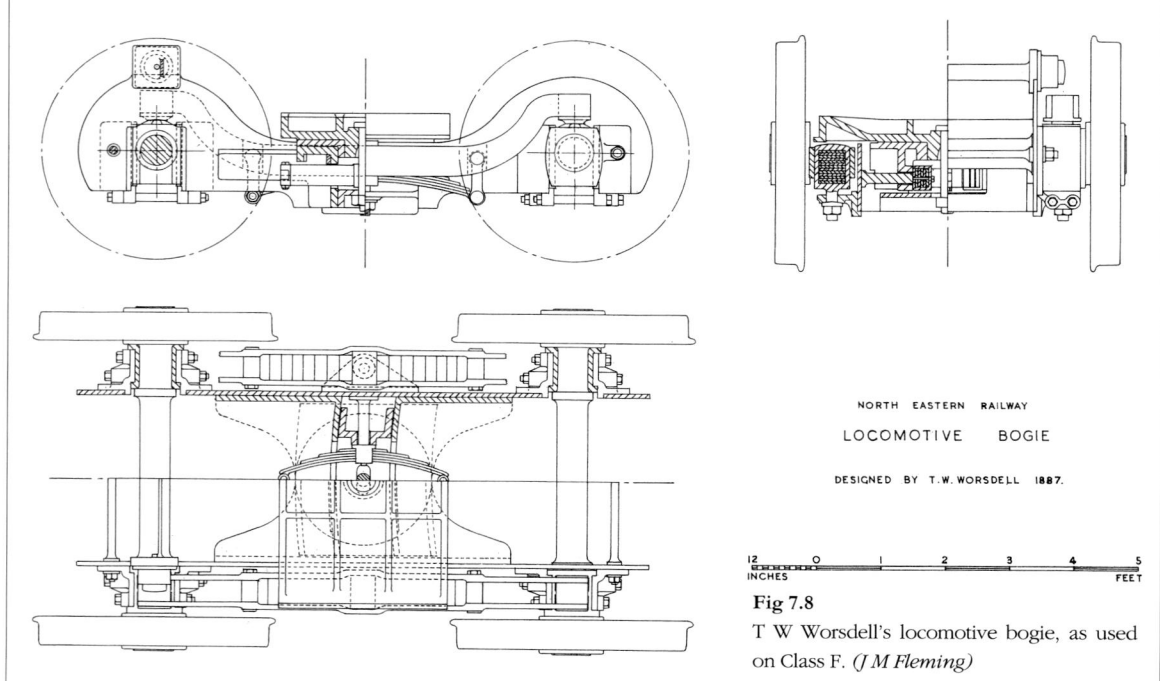

Fig 7.8
T W Worsdell's locomotive bogie, as used on Class F. *(J M Fleming)*

Fig 7.9
Class G1 No 274. *(J M Fleming/ traced J F Addyman)*

Plate 7.7 (top)
Class G1. No 328 in its original form.

Plate 7.8 (lower)
Class G1. From 1900 to 1904 the class were rebuilt as 4-4-0s. No 372 was rebuilt in June 1903 and later received the superheated boiler shown here. The provision of extra coal rails on the tender was common practice.

above the low pressure; the latter were actuated by a single valve rod. A 2in diameter spring-loaded relief valve was fitted at each end of the cylinders to allow any trapped water to escape.

The reason for the return to the 2-4-0 wheel arrangement is obscure. However, it is possible that a second pair of frames may have been prepared in 1886 for a simple version of the Class D compound for comparative trials and that, instead, they were used for the experiment with piston valves.

Worsdell Singles: Classes I and J

The invention of steam sanding apparatus, in the 1880s, brought locomotives with a single pair of driving wheels back into favour and resulted in the introduction of the Class I 4-2-2 compounds, ten of which were built in 1888-90 (Plates 7.10 & 7.11 and Fig 7.11). When new, the first five worked on the York to Newcastle main line, but were soon replaced by the larger Class J compounds. Class I had 7ft 1¼in driving wheels, with cylinders and motion similar in most

Plate 7.9
Class B compound No 855, showing the low-pressure cylinder and valve chest covers on the viewer's left and the high-pressure on the right.

respects to the existing compounds, a boiler common to Classes B and C, and a bogie the same as Classes F and F1. The last five locomotives had a pipe leading from the low-pressure exhaust to the tender through which some of the exhaust steam could be diverted, under the control of the driver, to heat the feed water. Any uncondensed steam could escape through a small chimney near the back of the tender.[2]

The Class J 4-2-2 compounds, with 7ft 7¼in driving wheels, were intended for the principal express passenger trains; ten were built in 1889-90 (Plates 7.12 & 7.13 and Fig 7.12). They had a high-pressure cylinder 20in by 24in and a low pressure cylinder 28in by 24in, placed at 2ft centres and with the steam chests outside the frames, the valve faces being vertical. To accommodate the cylinders between the frames the centre line of the high-pressure cylinder was inclined upwards and that of the low-pressure downwards towards the driving axle.

The valves were actuated by Joy's gear and rocking shafts (Fig 7.13). The rocking shafts occupied the positions where the sandboxes were normally situated, so that the latter had to be placed above the running plate where they were neatly combined with the driving wheel splashers.

The exhaust from the high-pressure cylinder entered a receiver pipe – 9in external diameter – that curved around the inside of the smokebox and came out at the right-hand side before passing down to the low pressure steam chest. The low pressure exhaust pipe likewise entered through the right-hand side of the smokebox to be joined to the blastpipe. The external portions of the receiver pipe and exhaust pipe were protected by a neat cover at the side of the smokebox, a similar cover at the opposite side of the locomotive preserving its symmetry. The front, back and top of these covers were painted black, but the sides were green, with a claret border and white line, the compound number plate being affixed in the centre.

---

[2] This system was also adopted on the last fifteen locomotives of Class F and Classes I, J, M and M1.

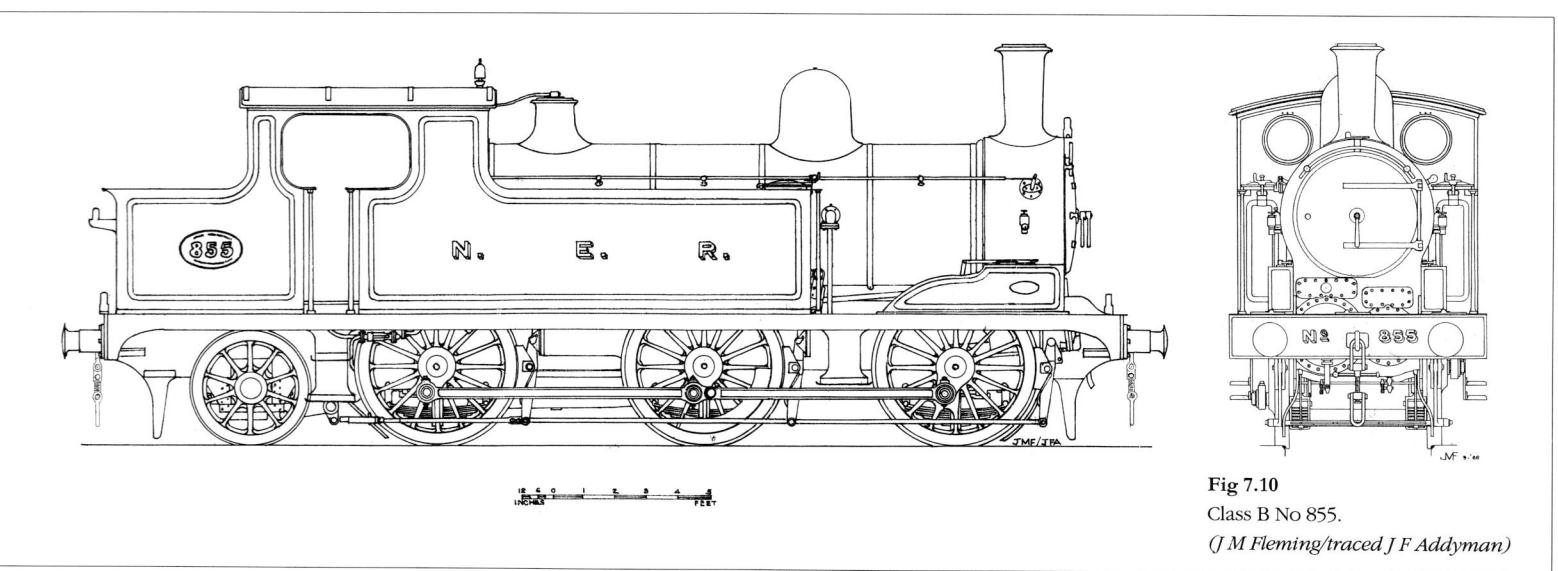

Fig 7.10
Class B No 855.
(J M Fleming/traced J F Addyman)

Plate 7.10
Class I. No 1326, newly-built and photographed in grey, with T W Worsdell's version of the armorial device on the splasher.

Plate 7.11
Class I. No 1329 rebuilt in 1900 as a simple and given outside bearings to the rear axle. This view was taken at the west end of platform 8 (in then numbering) at Newcastle Central station before 1906. Note the elevated ground signal, a feature of the station's signalling prior to 1909.

Fig 7.11
Class I No 1326 in its original form as a 2-cylinder compound. *(J M Fleming)*

Plate 7.12
Class J. No 1517 in its original condition but in Wilson Worsdell livery, with his small version of the NER armorial device (reproduced in *North Eastern Record Volume 2* - page 68) on the splasher. The low box between the splasher and cab is a sandbox for use when running tender first.

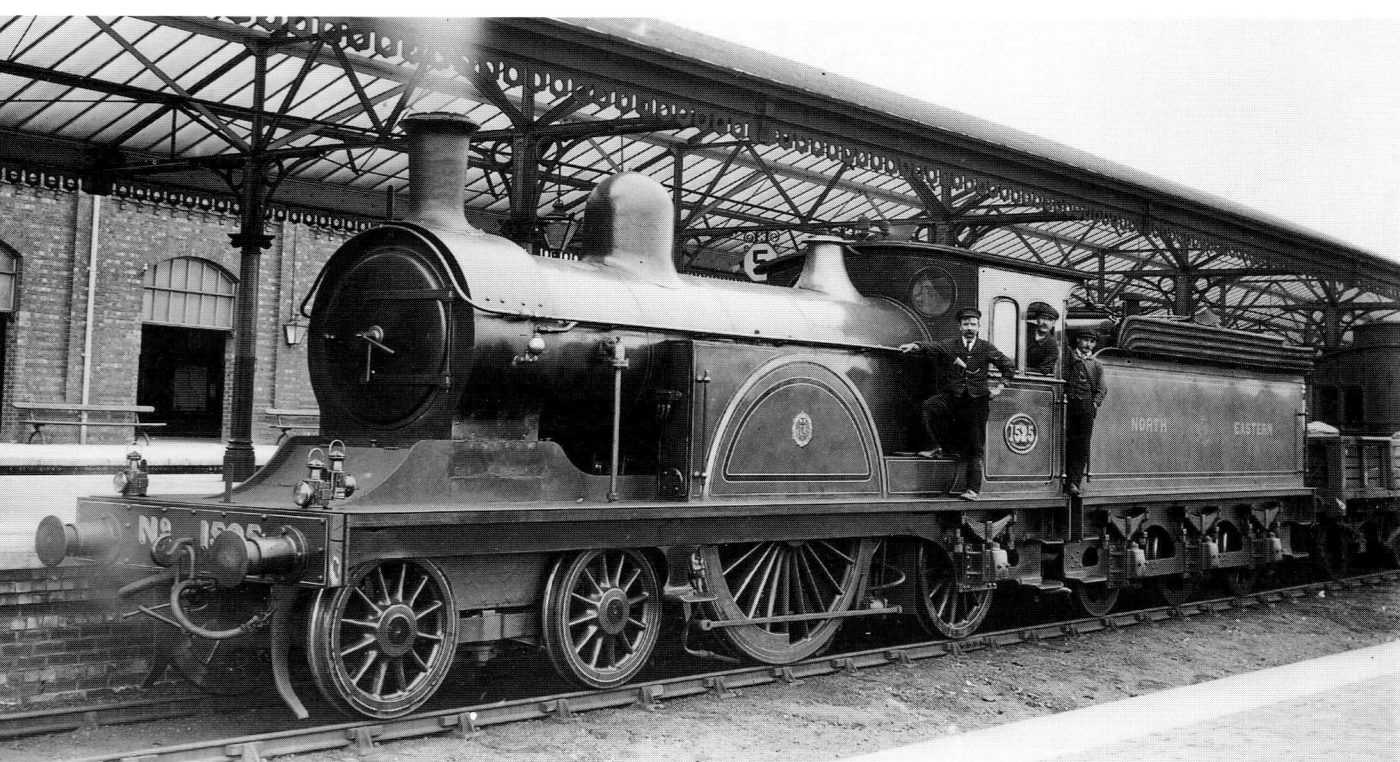

Plate 7.13
Class J. No 1525 rebuilt as a simple in 1896 and seen at Bridlington with a fish wagon marshalled in front of the passenger carriages.

Fig 7.12
Class J No 1526 in its original form as a 2-cylinder compound. (*J M Fleming*)

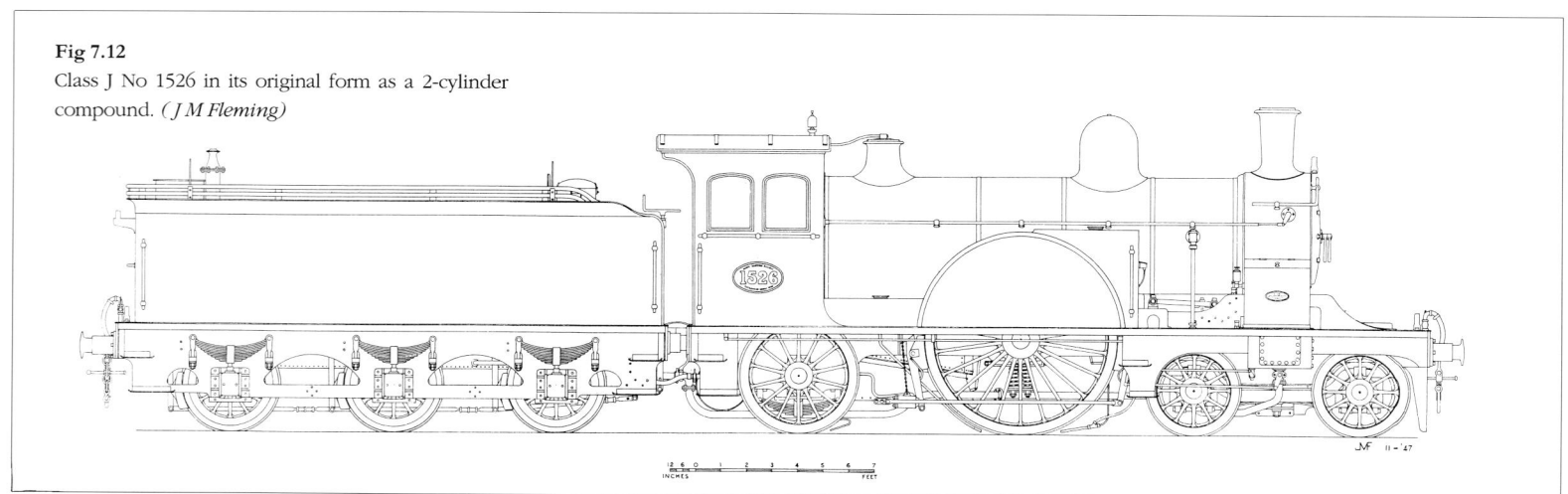

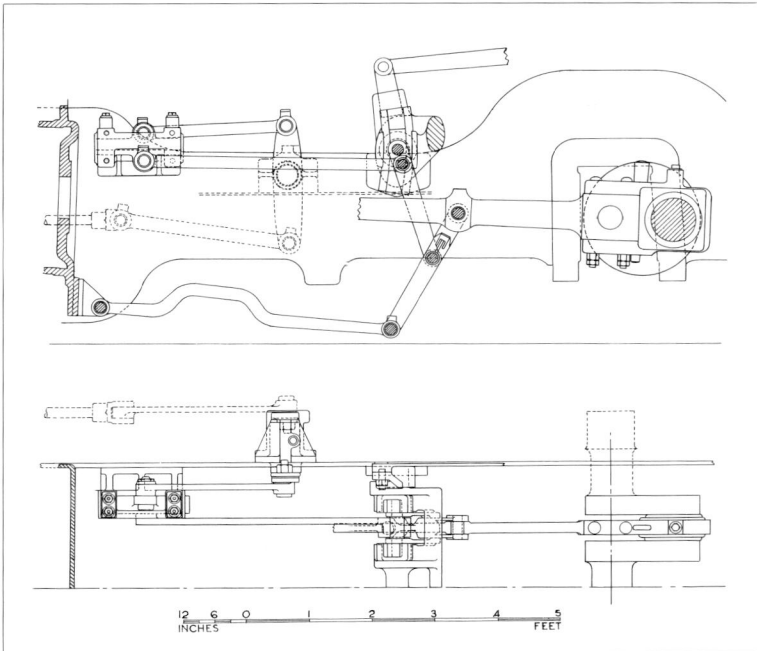

Fig 7.13
Class J - arrangement of Joy valve gear. The locomotive is shown in forward gear, and the high pressure piston is at its maximum travel of stroke in the back centre position. *(J M Fleming)*

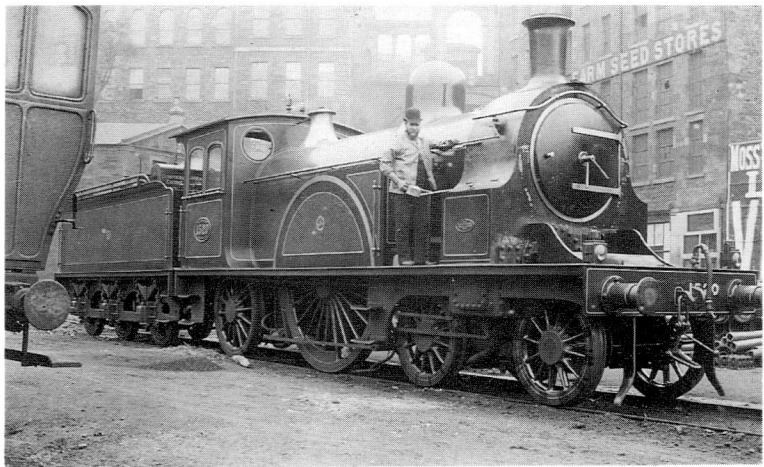

Plate 7.14
Class J No 1520, seen with driver William Smart at Edinburgh Waverley on 22 September 1890; Regent Bridge is in the background. *(J Braithwaite. J F Mallon collection)*

The splashers of the Class I and J singles had a 4in claret border and a black panel with ¼in white and vermilion lines; the company's heraldic device was displayed in the centre of the splashers. The sides of the sandboxes on Class J locomotives had a 4in claret border and a single ¼in white line.

The tenders built for the Class J locomotives had increased coal and water capacity, and all had equipment for heating the feed water. A short-lived custom, around 1889-90, was to have the driver's name and shed painted, in 2in letters, on one of the toolboxes, together with a basic description of the locomotive on the other. Several instances of this custom are to be seen in photographs of Class F and J locomotives, of which No 117 is an example:-

COMPOUND PASSENGER	ROB[T] NICHOLSON DRIVER
ENGINE N[O] 117	GATESHEAD

Class J No 1521 was shown at the International Exhibition of Science and Art that was held in Edinburgh in 1890 to commemorate the opening of the Forth Bridge. A correspondent of *The Engineer* singled out the NER compound for special praise, remarking that " . . . the finish of the engine is superb . . ."

Fig 7.14
Class J No 1524 rebuilt from compound to simple in June 1895. *(J M Fleming)*

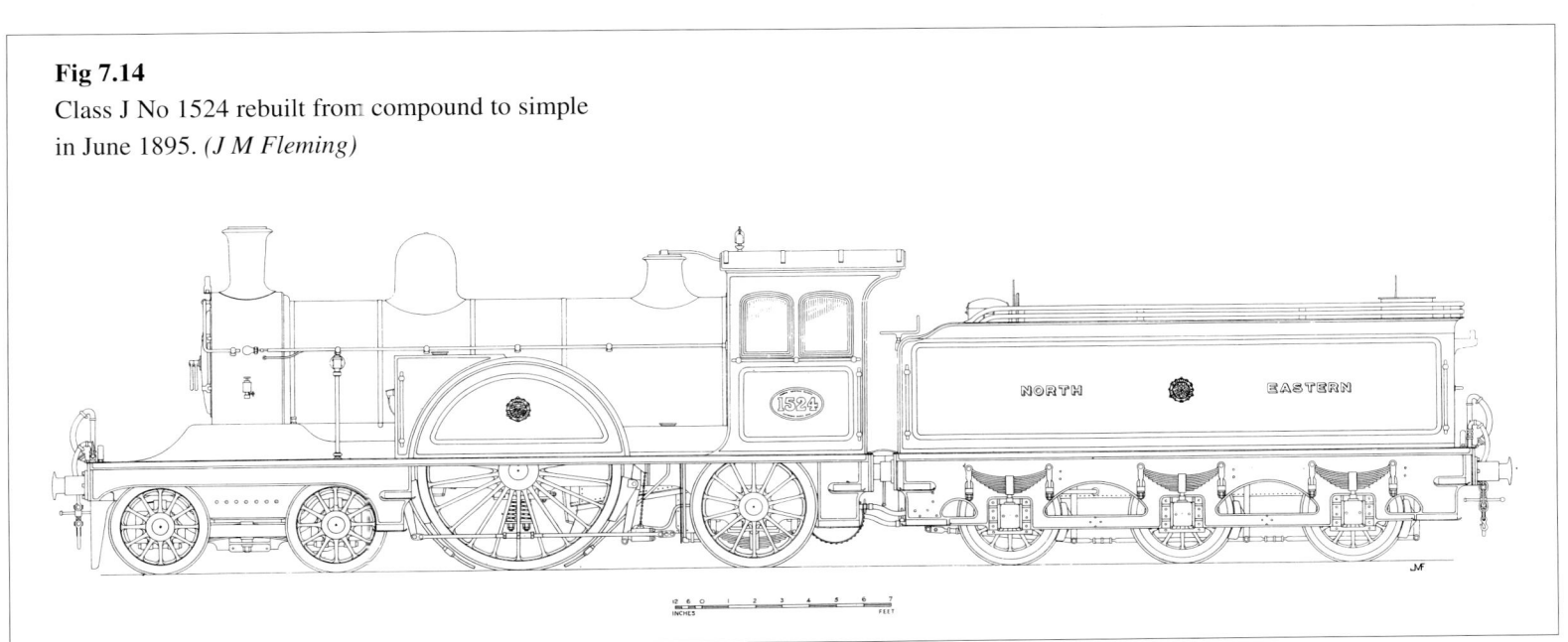

Plate 7.15
Class H. Probably No 1303, complete with shunting poles, shown in the later black livery. No 1310, now preserved on the Middleton Railway, has the rear stanchion of the cab considerably broadened by rivetting on an extra length of plate, a change which took place in NER days.

Fig 7.15
Class H, built at Gateshead in December 1888. It is shown in its original livery.
(J M Fleming)

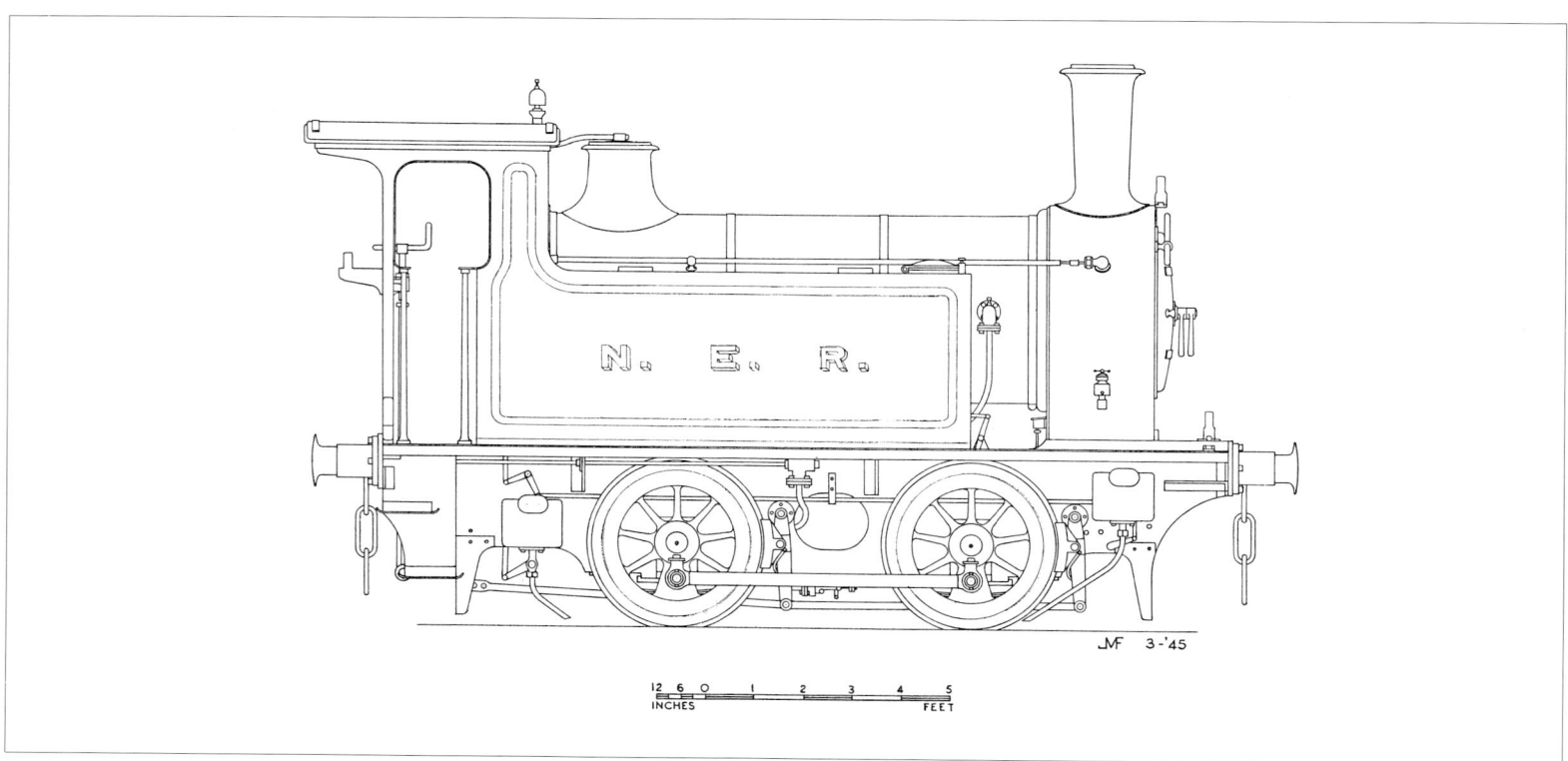

Fig 7.16
Class H1 crane tank No 995. *(J M Fleming/traced J F Addyman)*

Plate 7.16
Class H1. No 995, seen in its original condition. The crane was slewed round by a handwheel and screw. The load was raised by admitting steam to a cylinder, which drew in the rope and had a stopcock to retain the steam. The load was lowered by gravity, under the control of a lever-operated brake.

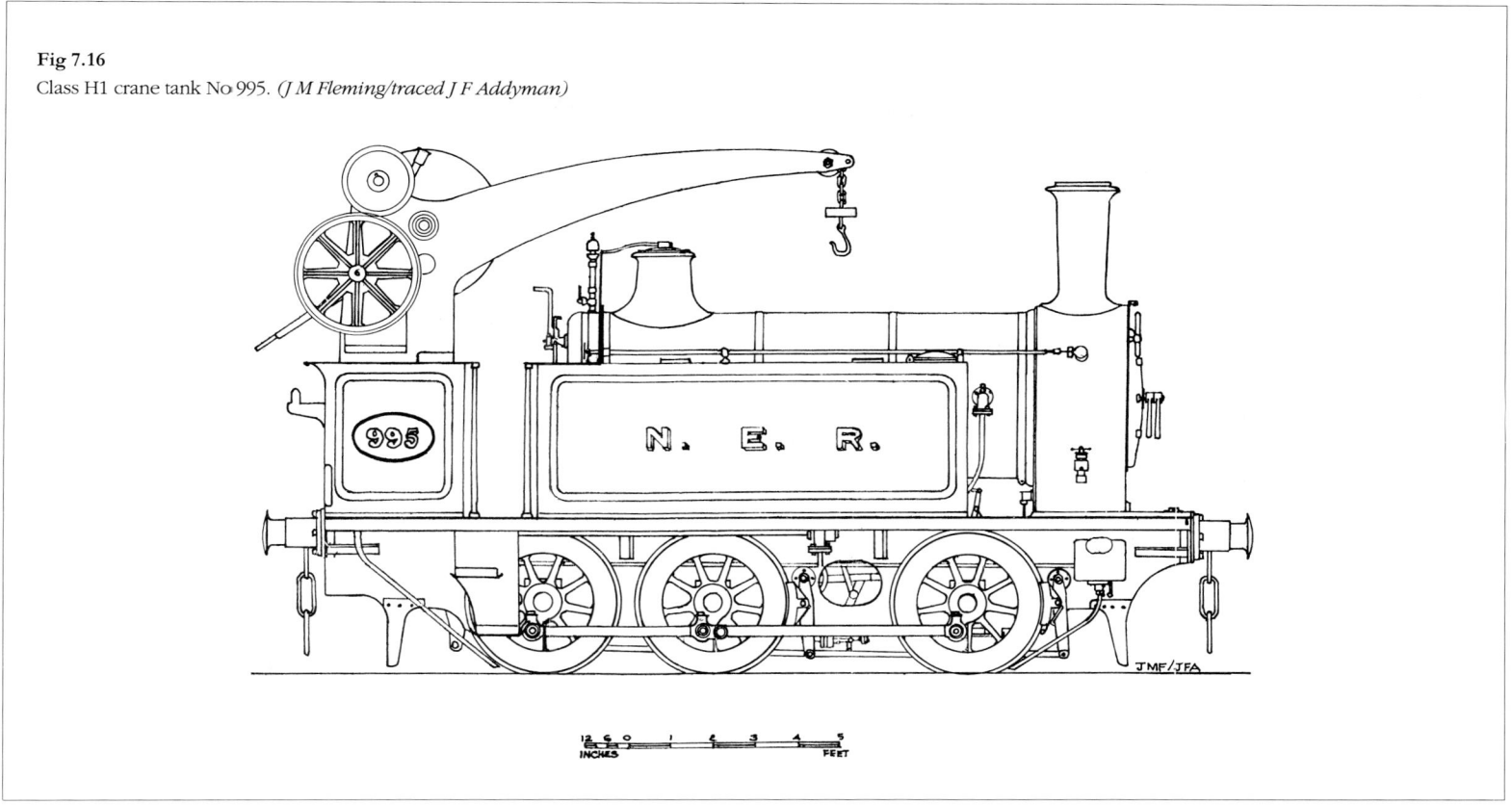

Plate 7.17
Class K.

Small Tank Locomotives: Classes H, H1 and K

Two small types of tank locomotives were built in 1888 for shunting around the railway workshops and in yards where there were tight curves, these being the Class H 0-4-0Ts (Plate 7.15 and Fig 7.15) and the Class H1 0-6-0 crane tanks (Plate 7.16 and Fig 7.16). The locomotive portions of the two classes were identical but Class H1 had extended frames to carry the three-ton steam crane. Coal was carried in a small bunker at the rear of the left-hand side tank. Class H1 carried brass numberplates on the side sheets around the crane platform, but Class H had the locomotive number in gold transfers, in lieu of a numberplate, in the centre of the back panel of the cab.

Class K was an even smaller 0-4-0T, designed for use at Hull docks where the maximum permissible axle weight was only 7½ tons in some places (Plate 7.17 and Fig 7.17). In fact, by cutting down weight wherever possible and incorporating a 'marine' type boiler with the firegrate inside a cylindrical flue, the weight in working order was kept down to 13 tons 14 cwt. The cab was reduced to little more than a weatherboard and a roof and, as with Class H, the number was in gold transfers on the back panel. The rear lamp bracket was on the roof support.

Class K was the last design introduced by Thomas Worsdell. In September 1890 he resigned owing to his failing health and was succeeded as Locomotive Superintendent by his half-brother, Wilson Worsdell.

Rebuilding

During T W Worsdell's short regime many of his predecessors' older locomotives were given new boilers and smokeboxes of the new standard designs, and the coal capacity of the tenders was increased by the provision of coal rails. The new style of boiler mountings, the elimination of the smokebox wingplates, from those locomotives which had been supplied or rebuilt with them, and, in many instances, the filling-in of the cutaway sides of the cabs made a considerable change in the appearance of the locomotives. New driving wheel splashers were often provided, or else the ornamental openings in existing splashers were filled in, the brass beading around these openings sometimes being left in position.

There was a wholesale removal of the makers' brass plates, very few of them – apart from those on the *Tennants* – surviving the clearance. Many of the Fletcher number plates were also replaced by Worsdell plates showing the locomotive number, the company's name, the works where the locomotive was built and the date of building. Where the Fletcher numberplates were retained, the initials of the maker, and the date, were die stamped below the number.

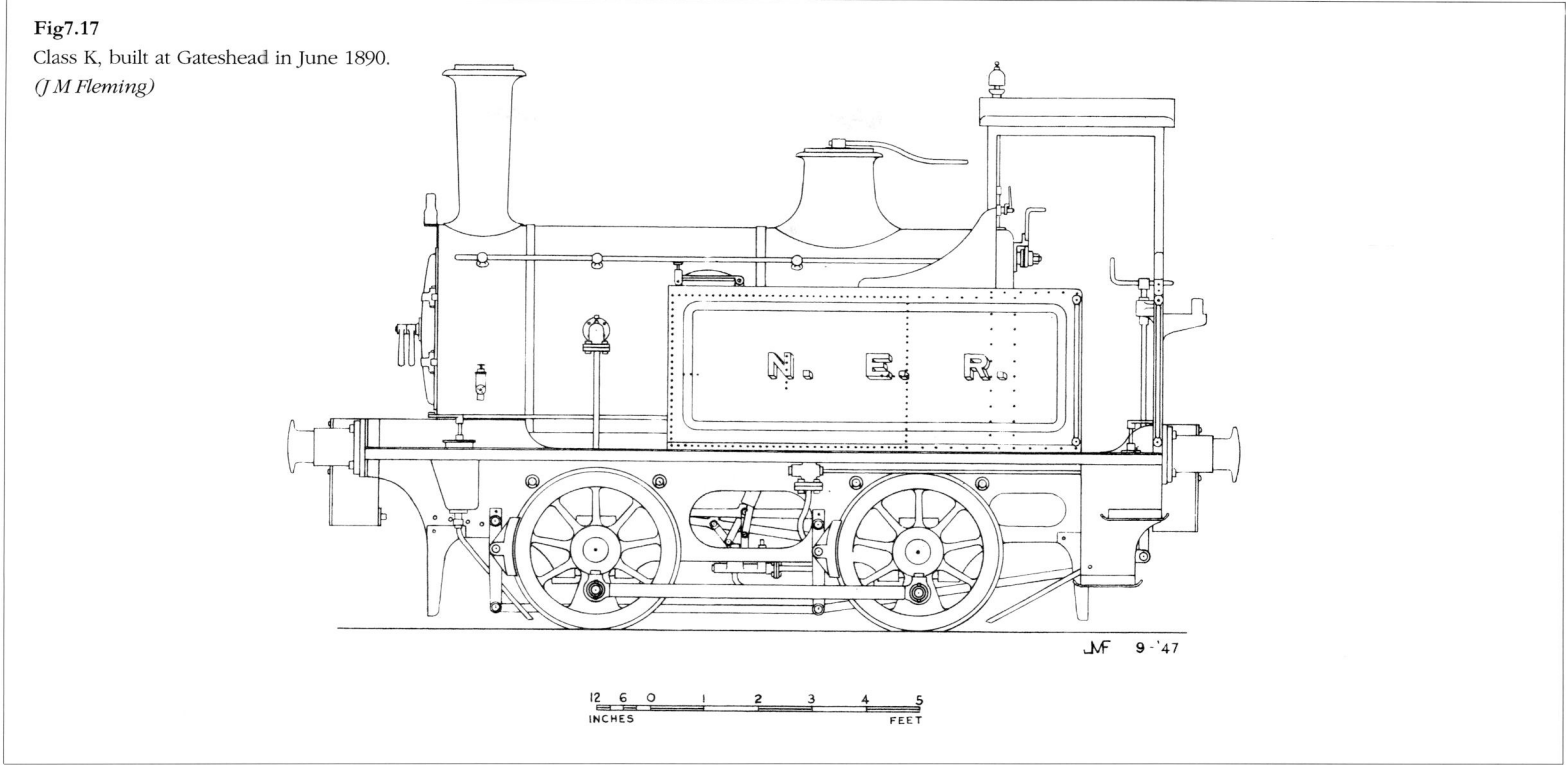

Fig 7.17
Class K, built at Gateshead in June 1890.
(J M Fleming)

Liveries

A few of Fletcher's goods locomotives appeared at first in the T W Worsdell livery without the initials 'N E R' on the tender. Another early variation was seen on the tender of McDonnell 0-6-0 No 78, which had small four-pointed gold stars equidistant between the initial letters 'N E R', and spiked hyphens, about 6in long, a similar distance in front of the 'N' and after the 'R'; the two portions of the cab sides of No 78, above and below the horizontal handrail, were separately panelled and lined out (for example see Plate 6.3). However, most of the McDonnell cabs, and those of the Fletcher locomotives, had a single black panel, with the usual white and vermilion lining, which followed the contours of the cab and extended up to about 6ft above the level of the running plate. Photographic evidence for the colour of the roof of Fletcher cabs, and for the colour of the top of saddletanks is inconclusive. They may have been claret colour overall, although one might expect the tops to be painted black for reasons of economy. The Fletcher bogie tank passenger locomotives and the various classes of saddle tanks did not carry the company's initials.

Even during the T W Worsdell era, and beyond, there were variations in the shades of the colours used by the different locomotive works. Gateshead used Saxony green, with vermilion for lining-out and for buffer beams; Darlington's shade of green is said to have had a blue tinge. and scarlet lake was used for lining-out and for buffer beams: York used a green colour with a yellow tinge – a vivid grass green – and orange vermilion for lining-out and buffer beams. There were variations, too, in the finish of the interior of the cabs. Gateshead cabs had simulated light-oak graining, with a black border and white line: Darlington generally had walnut graining, and York cabs varied in light and dark shades.

Plate 7.18
A T W Worsdell interior. The cab of a Class C 0-6-0. The fireman's handbrake is on the left; the driver's was on the tender, on the right-hand side. Above the left-hand side window is the visual indicator for Raven's electrical cab signalling system (much shorter-lived than his mechanical 'fog-signalling' system), with its bell above the right-hand side window and battery box below. Below the driver's spectacle window, left of the reversing lever, is the stopcock for the steam supply to the Westinghouse (brake) pump, with the driver's brake handle below it. The conspicuous handwheels flanking the firebox controlled the injectors.

Chapter 8

Locomotive Progress 1890 – 1910
Wilson Worsdell, Locomotive Superintendent

There were few signs of change during Wilson Worsdell's first two years as Locomotive Superintendent.[1] Existing orders for locomotives were completed and were followed by orders for further additions to Classes A, C, E and H. The Class C 0-6-0s built at Darlington in 1891 showed a minor variation from the original livery, the black panel band on the cab sides being extended upwards to surround the side windows.

The first of Wilson Worsdell's new designs to appear was the large Class L 0-6-0T, with 19in by 24in cylinders and Joy valve gear, and in unchanged T W Worsdell livery. Ten of these were built in 1891-92, and they retained their Joy gear to the end of their long existence (65-68 years) (Plate 8.1). Their 4ft 7¼in wheels introduced a NER standard that lasted through the Raven era to Class T3.

Passenger Locomotives 1892-94

In 1892 a more powerful class of passenger locomotives, the celebrated Class M1 4-4-0, was introduced for the main line expresses (Plate 8.2 and Fig 8.1). Twenty were built, together with one Class M two-cylinder compound of similar dimensions for comparison. Class M1 had 7ft 1¼in diameter coupled wheels and 19in by 26in cylinders, with the steam chests placed outside the frames and valves actuated by Stephenson link motion and rocking shafts; the final engine – No 1639 – had Smith's patent piston valves. The increase in the piston stroke, from 24in to 26in, and the reintroduction of link motion instead of Joy's radial gear were significant changes from T W Worsdell's practice.

The Class M compound had a 20in by 26in high pressure cylinder on the left-hand side, with its centre line inclined upwards towards the driving axle, and a 28in by 26in low pressure cylinder on the right, inclined downwards. As with the Class J 4-2-2s there was a cover over the external low pressure steam and exhaust pipes at the right-hand side of the smokebox and a similar cover on the opposite side for the sake of appearance. At the date of their introduction Classes M and M1 were the largest and heaviest express engines in Great Britain, weighing over 90 tons

---

[1] His post was redesignated Chief Mechanical Engineer in 1902.

Plate 8.1
Class L. No 553, photographed in grey. It is lined out in T W Worsdell style, up into the top corners of the cab sides.

Plate 8.2
Class M1. No 1638 in Wilson Worsdell's new livery which came into general use with this class.

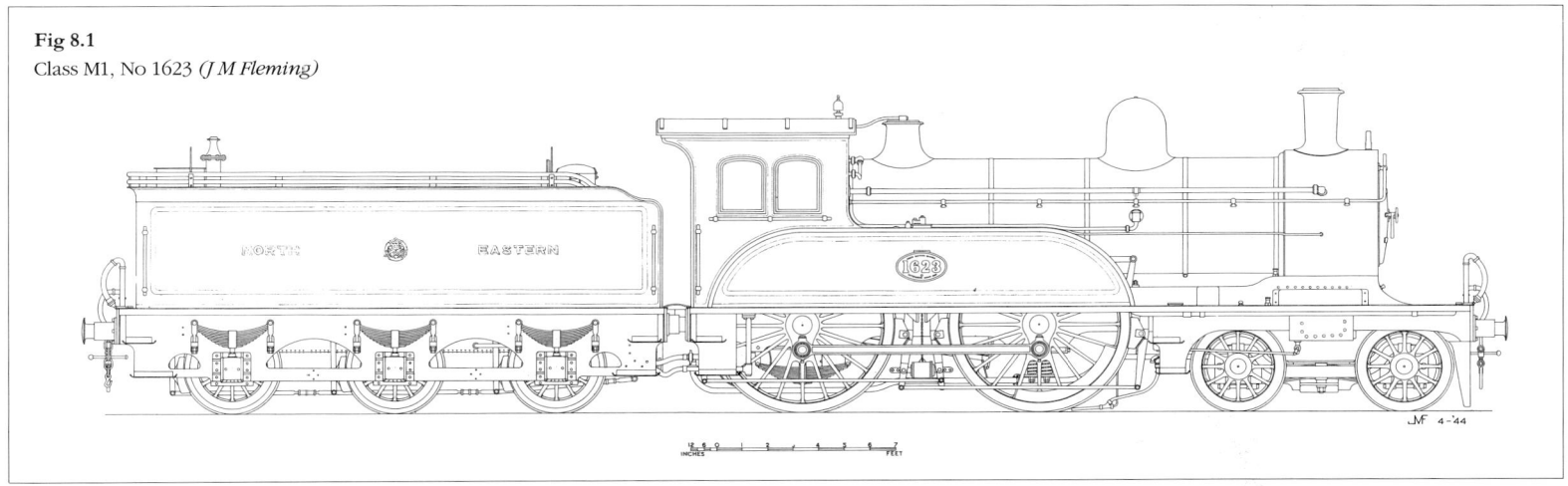

Fig 8.1
Class M1, No 1623 (*J M Fleming*)

Plate 8.3
Classes M and M1. No 1619, the compound, is distinguished from No 1623 of Class M1 by the large cover over its steam pipes.

No 1623 was chosen for the official photographs of Class M1 – probably because it was completed in May 1893, which was also the building date of Class M No 1619, the two engines, without their tenders, being posed together end-to-end at Jarrow Slake (Plate 8.3).

Wilson Worsdell's Standard Livery
A modified livery was introduced on the first engine of Class M1 – No 1620 – completed in December 1892, Saxony green being retained as the main colour but without the expensive claret colour for the borders etc. As before, boiler lagging belts were black, with a white line on each edge. The black panel bands on the splashers and tender had a $1/8$in white line on both sides; the 1¾in black border along their lower edges was also separated from the green paintwork by a $1/8$in white line. The cab sides had a 1inch black border and a $1/8$in white line at the front edge and above the splasher beading, the width of the border being increased to 2¼in at the rear edge to include the raised beading and to 1¾in around the window openings. The frames, both above and below the running plates, were black. The sides and back of the tender had 2¼in black borders and a $1/8$in white line at the front and along the upper edge of the curved flare at the top. A new form of the NER heraldic device, incorporating the company's name around the border, made its first appearance on the Class M and M1 tenders. The angle irons and footsteps below the running plates of the engine and tender, with the tender frames, were black with vermilion lining. The front and top surfaces of the tender tank, the coal space and the coal rails were black.

The new livery had been tried on at least one old engine before 1892, the engine concerned being a small 0-4-0ST, No 1781 in the duplicate list, which was withdrawn and sold by the NER in December 1891. However, from photographic evidence, it is clear that Wilson Worsdell's new livery was not brought into general use until December 1892. For example, Class 964 long-boiler 0-6-0 saddle tank No 964, reboilered in November 1892, was turned out in T W Worsdell's livery whilst No 967 of the same class, reboilered in December 1892, had the new livery. (On 1 January 1893 these engines were renumbered 1753 and 1756 respectively in the duplicate list, only to be restored to capital stock as Nos 1657 and 1659 a year later.)

Wilson Worsdell's new livery suited the locomotives of pre-1886 design very well. Splashers, cabs, sandboxes and tender tanks were green, with black borders and fine white lines. The sides of Fletcher's round-topped cabs were painted green but the roof colour is uncertain, it may have been black or a continuation of the green. In addition to the black borders, the tenders and cab sides had a 2in wide inner black panel edged with fine white lines. From December 1892 until about the beginning of 1894 the inner panels followed the contours of the cab side sheets but, from 1894, there was only a rectangular panel below the horizontal cab handrail. The black panels were omitted altogether from the cab sides of the *Tennant* locomotives.

Saddle tanks constructed with snap-head rivets – such as Class 1350 – were bordered and panelled in a similar manner to cabs, the upper surface being black. However, the countersunk-riveted saddle tanks of Class 964 had only a panel on the sides, the black border being omitted, and it seems likely that the green paintwork extended over the top of the tanks. Saddle tanks and engines with rear tanks – ie Class 124 0-6-0Ts and the 'Bogie Tank Passenger' engines – did not carry the initials 'N. E. R.'

Plate 8.4
Class N. No 1655 has been reboilered and appears in the later black livery, rather than its original green. From September 1905 the traffic classification was painted on the cab side below the numberplate; in this case it is K23. It was painted out again from 1910.

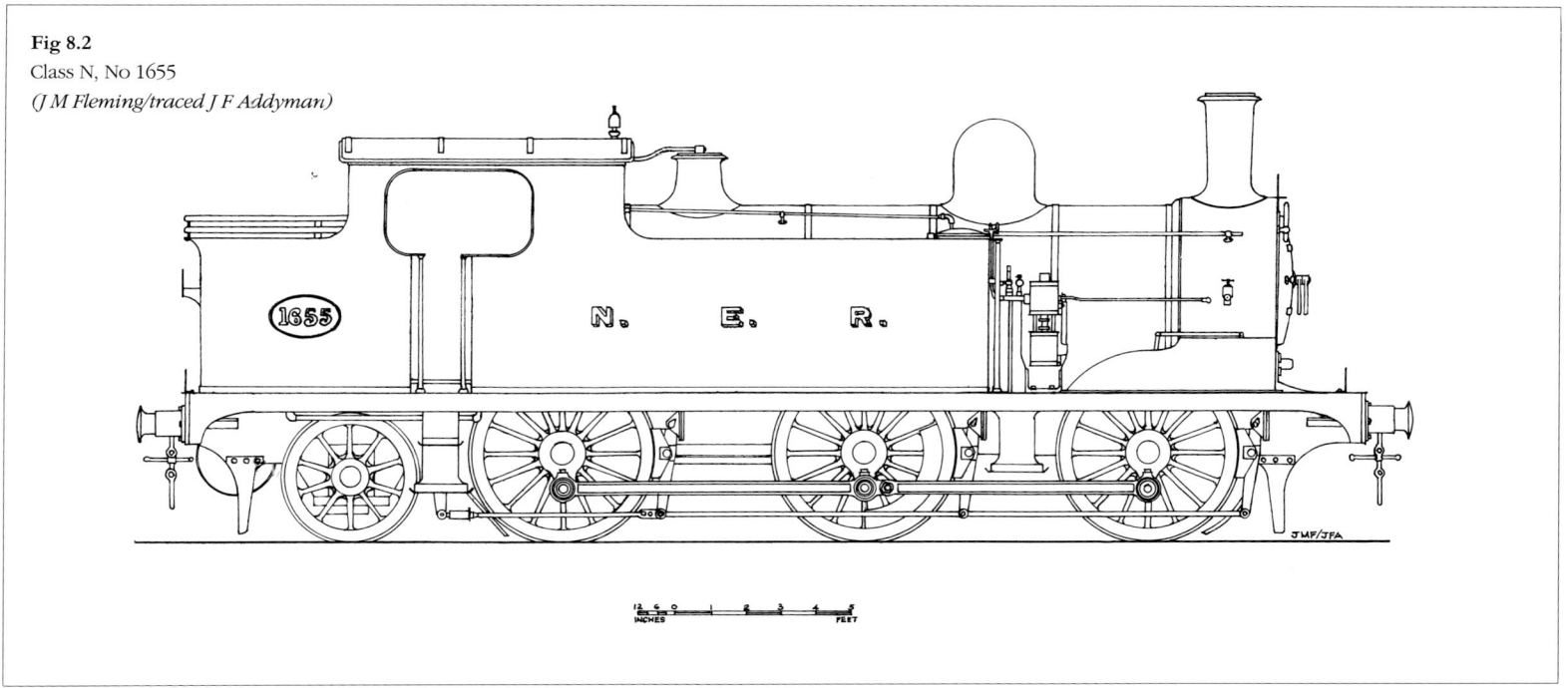

Fig 8.2
Class N, No 1655
(J M Fleming/traced J F Addyman)

Classes N and 964A

A new, more powerful class of 0-6-2Ts for local goods traffic – Class N – was introduced in 1893 (Plate 8.4 and Fig 8.2). They had 19in by 26in cylinders, with the valve chests placed above them and slide valves actuated by link motion and rocking shafts; the water capacity of the tanks was increased by a well tank below the bunker. Class N was similar in most other respects to Class B. An official photograph of Class N shows No 1655 – built in December 1893 – in plain 'shop grey' which gives no indication of the actual livery. However, it is known that they were the last new class to wear the T W Worsdell livery, and it seems likely that the original style of the panelling and lining-out on the tanks, cab and bunker would have been similar to that seen on Class 964A No 968 (Plate 8.5). This was a short-lived style that was soon superseded by the less elaborate panels and lining-out adopted for the Class O 0-4-4Ts in 1894.

Class 964A came about as a rebuilding of Fletcher's Class 964 long-boiler 0-6-0 saddle tanks (Fig 8.3). Of the 50 Class 964 locomotives built between 1872 and 1875, 15 were rebuilt during 1892-3 as

Plate 8.5
Class 964A. No 968, one of fifteen Fletcher 0-6-0ST of Class 964 rebuilt as side tank locomotives, using the same boiler as on Classes E and E1, during 1892-3. Seen in its original green livery. *(Locomotive Publishing Company)*

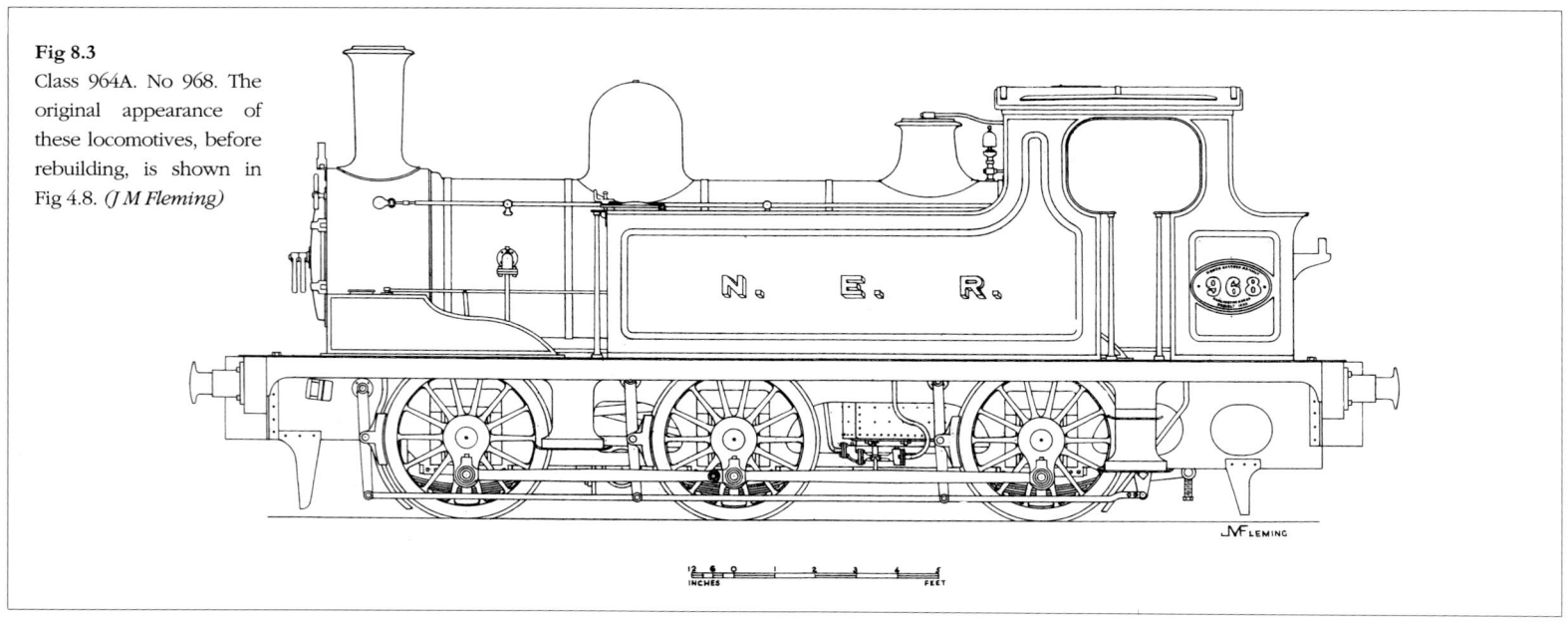

Fig 8.3
Class 964A. No 968. The original appearance of these locomotives, before rebuilding, is shown in Fig 4.8. *(J M Fleming)*

Plate 8.6
Class O. The bunker capacity of No 2087 has been increased by the addition of a cage to the original coal rails. On the right is the distinctive water tower of Neville Hill shed, Leeds. The Westinghouse pump, seen in front of the side tank, was originally housed at its rear in a cupboard within the cab. This had been standard practice on a number of classes but was changed because of the noise made by the pump.

Plate 8.7
Class O. No 2099 in its original condition.

Plate 8.8
Class O. Interior of the right-hand side of the cab, the driver's side on the NER. The combined lever and screw reverse was an unusual feature provided on all but the first 20 of Class O because of its mix of duties ranging from shunting to the working of local passenger trains, on which it was commonly seen. It provided the options of the rapid movement possible with lever reverse and fine control possible with the screw. A drawing of the lever and screw appears opposite page 371 of Warren's *A Century of Locomotive Building by Robert Stephenson & Co*, published in 1923. This photograph was taken during the British Railways' period, since the locomotive is 'push-and-pull' fitted.

conventional 0-6-0 side tanks with the wheelbase increased to 13ft 8in instead of their original 11ft 8in short-coupled wheelbase. All but two of the unrebuilt locomotives had been renumbered in the duplicate list by 1893 while, on rebuilding, the 15 locomotives of Class 964A were renumbered back into the capital list, taking numbers in the 800 and 900 series previously occupied by Class 964, though none of the rebuilds regained its original number.

1894-5: Classes O and P

The working of the Worsdell-von Borries compounds, compared with other classes of locomotives, came under scrutiny in 1893, the resulting report being unfavourable to the compounds. It followed that no more two-cylinder compounds were built and, in due course, all those already in service were converted to simples. When additional goods engines were required, the design selected was T W Worsdell's Class C1 simple 0-6-0, 20 of which were built in 1894-5.

Also in 1894 came the first of the 70 Class P 0-6-0s with 4ft 7¼in diameter wheels, together with the first of 110 Class O 0-4-4Ts with 5ft 1¼in diameter coupled wheels. Many details were common to both classes, including a standard boiler similar to Class A, 18in by 24in cylinders with the valve chests between them, and valves actuated by link motion (Plates 8.6 to 8.9).

Plate 8.9
Class P. No 1823 photographed in grey, though the lining corresponds to that of the working livery. These had handbrakes only when built.

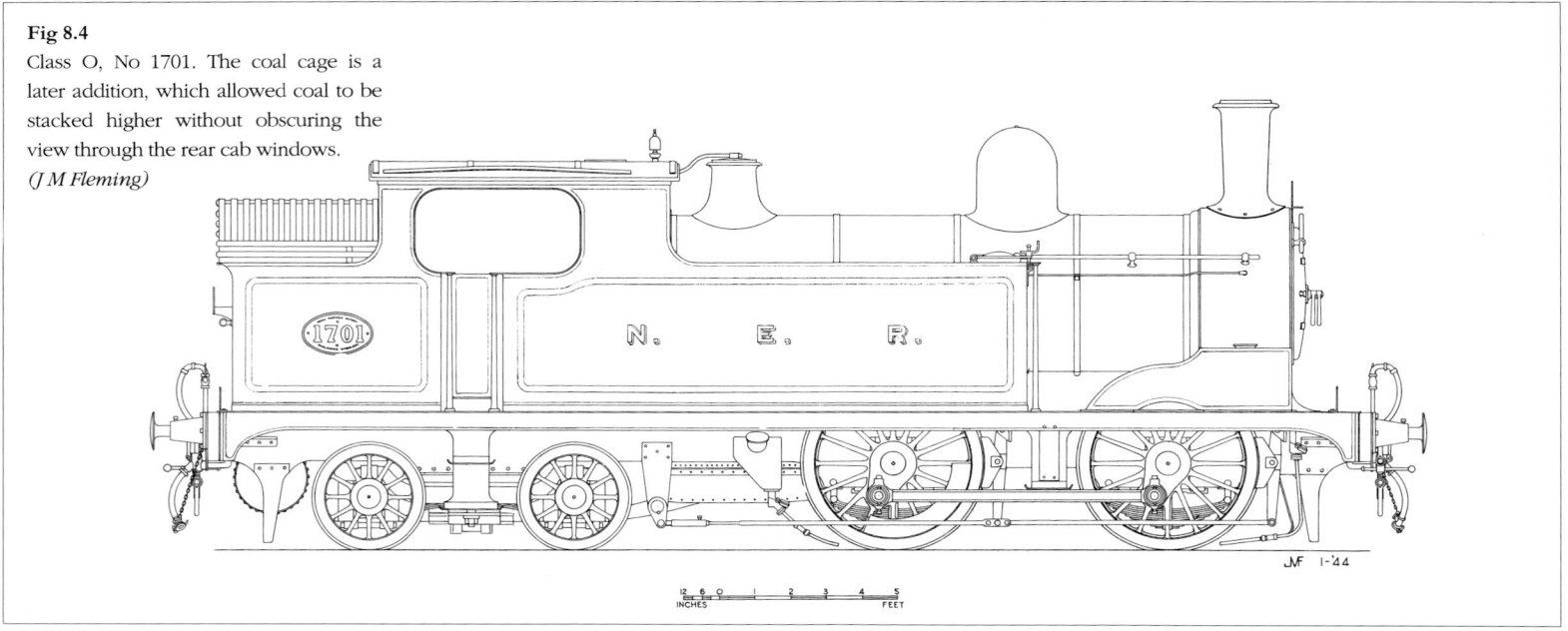

Fig 8.4
Class O, No 1701. The coal cage is a later addition, which allowed coal to be stacked higher without obscuring the view through the rear cab windows.
(J M Fleming)

The original livery and lining out of these Class C1 and P 0-6-0s was generally similar to that of Class M1, except that the tenders were lettered 'N. E. R.' The sides and tops of the splashers and front sandboxes had a ¾in black border and a fine white line.

The side tanks and bunkers of the Class O engines had a ¾in black border and a fine white line along their upper edge; the width of the black border along the lower edge of the tanks and the upper edge of the cab sides and ends was increased to 2¾in to include the external 2in by 2in angle irons. The beading around the cabside opening was black, but without a white line. (In later years a black border with fine white line was painted around the cabside opening, and this sometimes extended down each side of the doorway.) The front and side of the tanks, and the sides and back of the bunker, had an inner black panel with a fine white line on each edge; the panel on the side of the tanks did not follow the contour of the cab side sheets as in Wilson Worsdell's earliest livery (Fig 8.4), but more nearly formed a rectangle. The last 40 engines of Class O, built in 1900-01, had cab doors of varnished wood - similar to those still to be seen on the preserved NER 2-2-4T *Aerolite* - but steel cab doors were later provided for the entire class. The steel doors were painted green, with black borders and white lining.

Compounds Simplified

The first compounds to be rebuilt as simples were the Class J 4-2-2s (Fig 7.14), the exposed position of their outside steam chests having been the cause of trouble due to cracking. Between January 1895 and October 1896 they were given 19in by 24in cylinders, with valve chests above, and 8in diameter outside-admission piston valves actuated by direct driven link motion. The feedwater heating equipment was removed and, from about May 1896, outside frames and bearings were provided for the trailing axle. The two Class D 2-4-0 compounds were also rebuilt, as Class F1 4-4-0 simples, in October 1896. The Class D engines and most, if not all, of Class J had already appeared in Wilson Worsdell's livery while they were still compounds; Class D No 1324 and Class J No 1521 even retained the burnished chimney top and wheel tyres, which they had as 'Exhibition' engines in 1887 and 1890, respectively.

The Class M two-cylinder compound – No 1619 – was rebuilt in 1898 as a three-cylinder compound, on W M Smith's system, with a single high pressure cylinder between the frames and a pair of low pressure cylinders outside. It was also given a larger boiler and, as a non-standard design, it remained the only three-cylinder compound engine on the NER. Although the combined splashers over the coupled wheels were retained, each wheel was separately outlined by a brass beading, these being edged with a black border and fine white line. Initially, each wheel arch also had an inner black panel, edged white, but this extra embellishment was short-lived (Plate 8.10 and Fig 8.5). The locomotive was classified 3CC.[2]

Classes Q and H2

Two new classes of express passenger locomotive appeared in 1896. The Class Q 4-4-0, 30 of which were built in 1896-97, was a development of Class M1 with 19½in diameter cylinders; the steam chests were above the cylinders instead of at the sides, the valves being actuated by link motion and rocking shafts. Although the boiler shell was the same size as that of Class M1, there were only 201 tubes instead of 225. The most obvious external changes were the shorter smokebox surmounted by a chimney with a polished brass top, and the raised clerestory roof on the cab. (Clerestory roofs had recently appeared on NER and East Coast Joint Stock carriages). (Plate 8.11)

---

2 For a time it had water-tubes in the firebox, an experiment also conducted on Class Q No 1929.

Plate 8.10
Class 3CC. No 1619 at Neville Hill.

Anticipating a resumption of very fast running in 1896 in competition with the West Coast companies, two Class Q1 4-4-0s were built with 7ft 7¼in coupled wheels and 20in by 26in cylinders. In addition to having larger coupled wheels than Class Q they had a longer coupled wheelbase and a longer firebox. They also had extra large number plates to suit the proportions of the very large splashers.

The painting and lining out of Classes Q and Q1 was similar to that of Class M1. However, from about 1896, engine numbers began to appear on the buffer beam directly above the drawhook. The previous style of 4in numerals was retained,[3] but without the 'No' prefix; the small black panel around the drawhook was also omitted. Not long after 1897 the numbers were re-positioned at the left-hand side of the drawhook.

Two 0-6-0 tank locomotives of Class H2 appeared in 1897, they carried the numbers 407 and 1787 (Plate 8.12 and Fig 8.6). The design was based on the Class H 0-4-0 tank and the Class H1 0-6-0 crane tank, but the 11ft 0in wheelbase was one foot longer than that of Class H1. The side tanks carried a total of 475 gallons of water,

---

[3] With a change in the style of shading.

Fig 8.5
Class 3CC, No 1619
(J M Fleming)

Plate 8.11
Class Q. No 1877. Class Q1 looked very similar.

though for working the Cawood Wistow & Selby Light Railway, acquired by the NER in 1901, No 407 was later fitted with longer and deeper tanks to increase the water capacity (Plate 8.13). A third locomotive, No 1662, with small tanks was built in 1907.

New Classes 1898-1902: P1, U, E1, and 290

The years 1898 to 1902 saw the introduction of no less than eight new classes of locomotives. In May 1898 the first of the Class P1 0-6-0s appeared, having the same size of wheels as Class P, but with a longer boiler and wheelbase the same as Classes C and C1. The frames were spaced 4ft 1½in apart, instead of the previous standard 4ft, and the cylinders were 18¼in by 26in with the valve chests between them. (The final 20 had 18½in by 26in cylinders.) Class P1 ultimately

Plate 8.12
Class H2. No 1787 originally appeared in green livery but is shown here in black NER livery, at Middlesbrough shed in 1925.

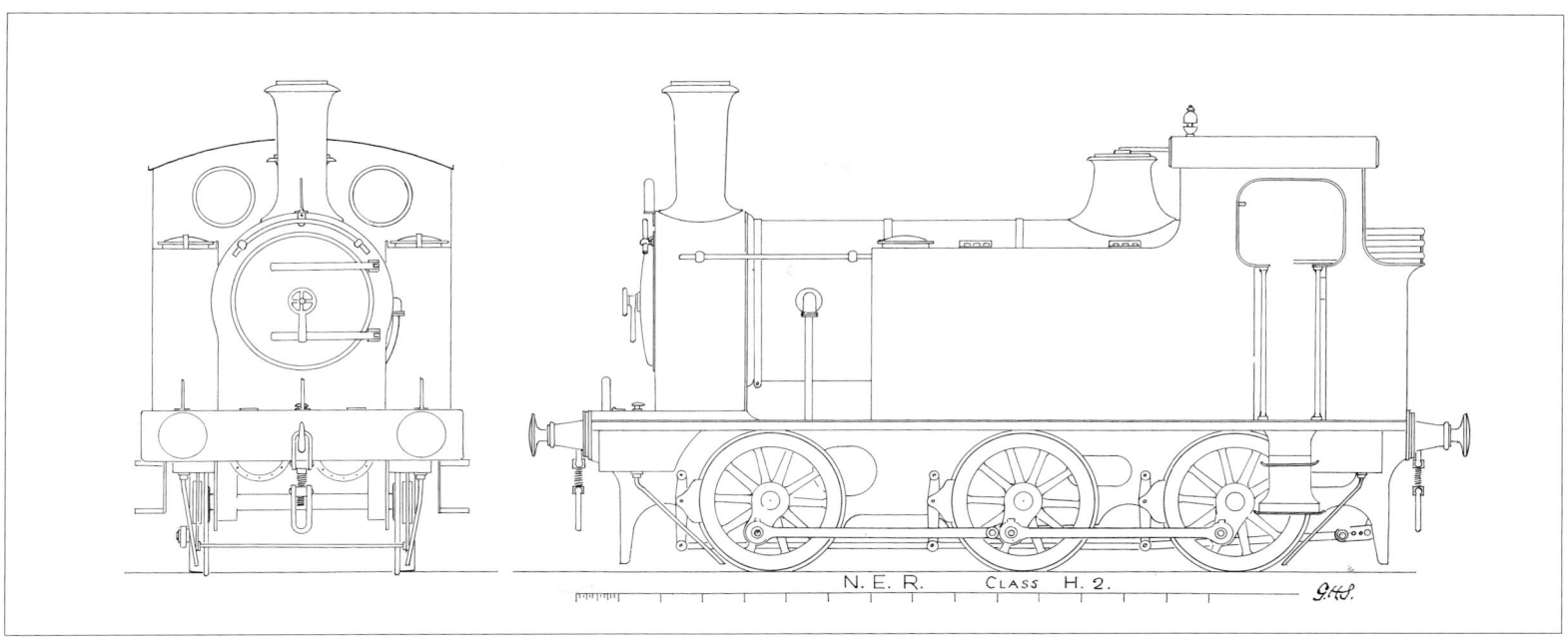

Fig 8.6
Class H2, built at Gateshead in March 1897. The coal rails are a later addition. A Westinghouse brake pump was provided (in the cab) for carriage shunting and soon proved useful for working the Cawood Branch. *(G H Swann)*

numbered 120 engines, built between 1898 and 1902, and had a tank engine counterpart in the Class U 0-6-2Ts, 20 of which were built in 1902-03 (Plate 8.14). Class U was the last new class to incorporate a standard 4ft 3in diameter boiler. Steel cab doors were provided on Class U from new. The original liveries of Classes P1 and U were similar to those of Classes P and O respectively.

The Class E1 0-6-0Ts, introduced in December 1898, were similar in many respects to Class E, but had 4ft 1¼in diameter wheels and 17in by 24in cylinders (Plate 8.15 and Fig 8.7a). Twenty were built in 1898-99, but then no more were added until 1914. In the meantime, fifty surplus 'Bogie Tank Passenger' 0-4-4Ts were rebuilt as Class 290 0-6-0T shunting engines, 40 being rebuilt at York Works between 1899 and 1904, and 10 at Darlington in 1907-08 (Plate 8.16). No 290 – one of the hybrid Fletcher/McDonnell engines of 1884 – was the first one to be altered, in June 1899. These rebuilds retained much existing material, including the front portion of the frames and the motion; the existing boilers and cylinders were also incorporated if still serviceable. New rear sections were fire welded to the old frames to accommodate the new trailing axle, and new 4ft 1¼in diameter wheels and new side tanks were provided.

Class E1 and about the first 35 engines of Class 290 had a green livery at first, similar to Classes O and U. However, there were

Plate 8.13
Class H2. No 407, in green livery, at Cawood on the Cawood, Wistow & Selby Light Railway.
(Lens of Sutton)

117

Plate 8.14
Class U. No 1667, seen in grey, lined in accordance with its green Wilson Worsdell running livery.

Plate 8.15
Class E1. No 1720 originally appeared in green livery but is seen here in the later black livery, with the red lining barely visible.

Plate 8.16
Class 290. These rebuilds of Fletcher BTP 0-4-4Ts as 0-6-0Ts varied considerable in appearance, depending on the proportion of the original locomotive retained. No 1346, built by R & W Hawthorn in 1875, was rebuilt at York in 1902 but kept its characteristic Fletcher cab.

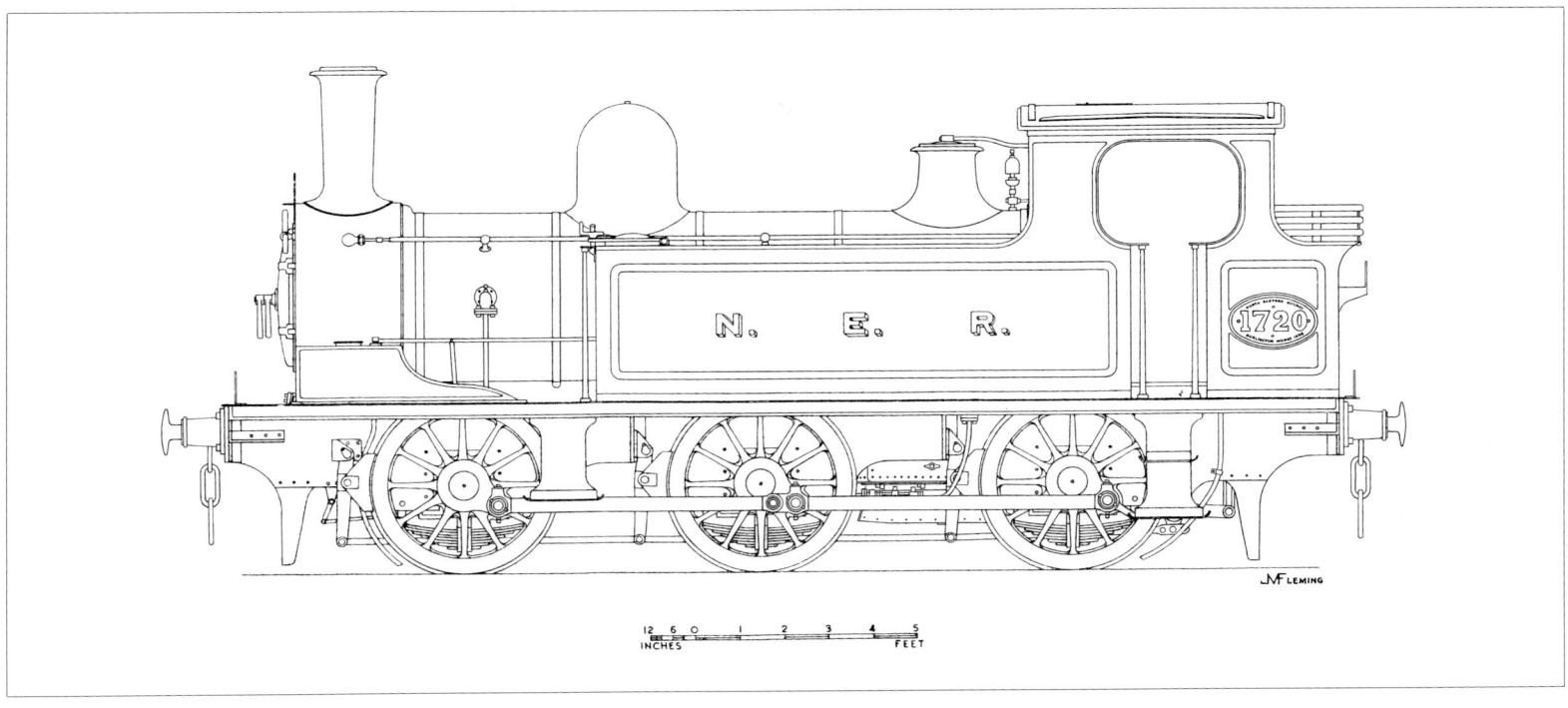

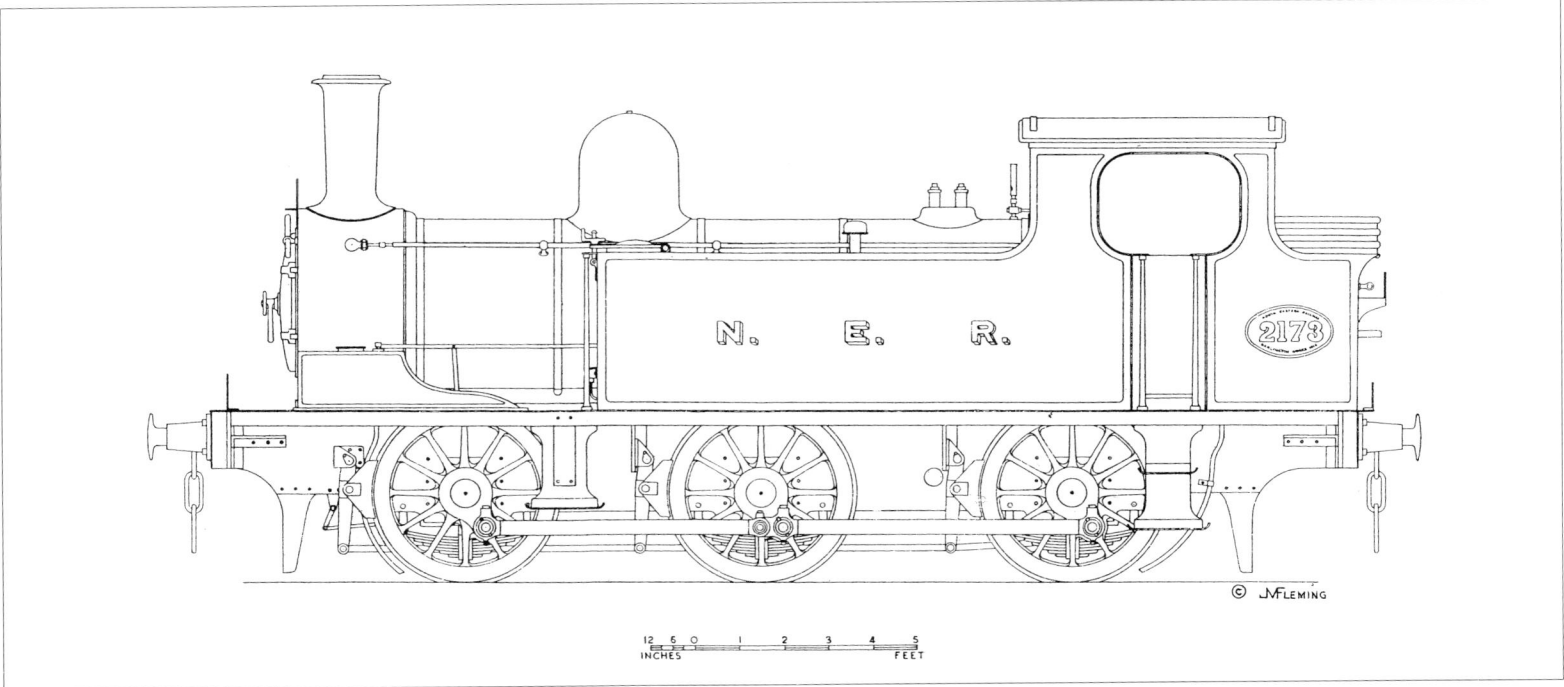

Fig 8.7a (top)
Class E1, No 1720, built at Darlington in 1899. *(J M Fleming)*

Fig 8.7b (bottom)
Class E1. Vincent Raven's revised design with large bunker, Ross pop safety valves, a different whistle and a vent pipe to the tanks (see chapter 9). No 2173 was built at Darlington in 1914. *(J M Fleming)*

variations in the style of the black panel on the bunkers of Class E1. As first turned out from Darlington Works, there was a gap between the vertical sides of the panel and the numberplate, but on some engines the numberplate was surrounded by a narrow black border and white line which merged with the sides of the panel. Similar variations were also to be seen on Class E and were possibly the result of individual preferences at the different paintshops.

Classes R, S and 4CC

The continuing increase in the weight of passenger trains – especially those composed of East Coast Joint Stock carriages – brought about the introduction of two new classes of locomotives in 1899. The Class S locomotives, which appeared in June, were the first 4-6-0s in the British Isles specifically designed for passenger traffic. They had 6ft 1¼in diameter coupled wheels – the driving pair having flangeless tyres – and 20in by 26in outside cylinders; the majority of the first ten locomotives originally had slide valves, but all eventually had piston valves. In contrast, the Class R 4-4-0s, which came out in August, had 6ft 10in diameter coupled wheels and 19in by 26in inside cylinders, with outside admission piston valves situated below them and directly driven by the link motion. Both classes had 4ft 9in diameter boilers with a working pressure of 200psi.

Class R was the North Eastern's most successful 4-4-0 design, being both reliable and economical. As an experiment, No 2011, the first of the class, was worked intensively for more than two years with two regular crews, covering 455 miles daily, six days a week. Between August 1899 and its first general overhaul in March 1903, No 2011 ran 284,182 miles (Plates 8.17 & 8.18).

Plate 8.17

Class R. No 2018, seen at York in its original form on a train of Lancashire & Yorkshire Railway stock. Its original Westinghouse pump barely showed above the top of the splasher (it stood between the coupled wheels) but in this view it had been replaced by the larger pump introduced about 1909 to cope with heavier East Coast trains. Class R tenders had 4ft wheels instead of the 3ft 9¼in standard, and this one has had its coal rails increased from two to four. Its bears the large armorial device which was reproduced in *North Eastern Record* Volume 1.

Plate 8.18

Class R. No 2011, built in August 1899 but seen with the superheated boiler and extended smokebox which it received in October 1917. The entire class received this treatment from 1912 onwards, while a number were also rebuilt with deeper Raven frames distinguished by having a convex profile at the front, rather than the concave outline of the two examples illustrated here.

Class R was the last 4-4-0 design to have the Worsdell combined splashers, but these had separate brass beadings outlining each wheel. The splashers had the usual black borders, with a white line, but the inner black panel seen on the previous 4-4-0s was omitted. A larger form of the NER heraldic device, with an ornamental border and having the company's name shown on a 'ribbon' around the base, was introduced on No 2015 in October 1899. However, later engines of Class R were still being turned out with the old device up to December 1899.

No 2110 hauled the royal train from York to Newcastle on 7 July 1906 when King Edward VII travelled to Tyneside to open the new bridge which bears his name. No 2110 had the older (1892) heraldic device on each splasher, and was decorated with flags and the royal coat of arms in front of the smokebox; the roof of the cab was painted white.

thereby reducing the total wheelbase to 48ft $4^3/_8$in so that the engines could be turned on a 50ft turntable; they also had short cabs with only one window in each side. Later engines with longer frames had normal cabs with two windows in each side, and the first three were soon altered to conform with the rest of the class. The standard 4ft spacing between the frames was tapered down to 3ft 10in alongside the cylinders and smokebox, and then by a further stage to 3ft $7^1/_2$in at the front buffer beam to allow extra sideplay for the bogie wheels.

The coupled wheels had separate brass-edged splashers, linked by sections of the coupling rod splashers, the latter extending forward to the rear of the cylinders to give a pleasing appearance. The cylinder lagging was painted green, with a black border and white line; there was also a brass beading at each end and a burnished steel outer cover at the front. The polished brass chimney cap, the collar at the front of

Plate 8.19
Class S. No 2003 was built with a shorter cab having a single side window, replaced in June 1901 by the one shown here. It was the only one of Class S to have Younghusband valve gear and the cranked reversing rod seen here. Class S1 was very similar in appearance.

Class R numbered 60 locomotives, 30 built in 1899-1901 and another 30 in 1906-07. The class outlived all the later NER express passenger locomotives, the last three being withdrawn in November 1957.

The Class S 4-6-0s created considerable interest when they first appeared (plate 8.19). The first three – Nos 2001-3 – had frames shortened by 2ft at the rear and tenders 8in shorter than standard,

the boiler and the safety valve cover, together with the burnished steel hinge straps on the smokebox door, added to the handsome appearance of Class S.

Official photographs of Nos 2001 and 2002 in original condition show the 1892 version of the NER heraldic device on the tenders, and it seems likely that 2003 would have been similar when new. However, an early photograph of No 2003, taken in 1901, before the cab was altered in June of that year, shows the new heraldic device – first used on Class R – on the tender and, in addition, the earlier device on the driving wheel splasher. This arrangement was adopted for the entire class and is seen in the official photographs of 'No 2006' (The subject of the photograph was actually No 2005, but the negative was retouched to read '2006'). The actual No 2006 was sent to the International Exhibition in Paris, in 1901, and was awarded the Grand Prix and a gold medal. Replicas of the medal were affixed to the driving wheel splashers after the locomotive returned to the NER, and the rear splashers carried a brass plate recording the Paris award; the NER heraldic device was re-located on the front splashers.

Nos 2009 and 2010 were lined in gold and decorated with flags and the royal coat of arms when they were rostered to haul the royal train conveying the Prince of Wales in June 1900. Gold lining was subsequently adopted for other principal classes of locomotives, and for those tank engines reserved for hauling the NER officers' saloons.

The gold lining was used either as a supplement to white lines, or as an alternative. Thus, boiler cleading belts were painted black, with a fine white line along each edge as usual, and an additional ³⁄₈in gold line each side of the cleading. On the tenders, however, although there was the usual white line on the outer edge of the black panels, gold was used instead of white for the inner line. Buffer beams were vermilion, with a black border and a gold line; there was also a narrow black border and gold line surrounding the base of the buffer casings. Buffer casings were vermilion, with a broad black border at

An enlarged version of Class S made its appearance in December 1900, and was designated Class S1. This had 6ft 8¼in diameter coupled wheels and a longer boiler than Class S, but the firebox dimensions were unchanged. The Ramsbottom safety valves were larger than those used hitherto, being 4in diameter instead of 3in, and they had a taller brass cover. Similar safety valves were fitted to the 4ft 9in diameter boilers provided for subsequent new classes and as replacements for the original safety valves on Class S.

The livery and fine finish of Class S1 was similar to that of Class S. The upper part of the frames between the coupled wheel splashers was painted green, with a black border and fine white line along the lower edge. Class S1 numbered only five locomotives – Nos 2111-5, built between December 1900 and August 1901 – after which the NER reverted to four-coupled designs for hauling its express passenger trains.

In June 1905 authorisation was given to build two four-cylinder compound 4-4-2s to the design of Walter Mackersie Smith, Wilson Worsdell's Chief Draughtsman. They were completed in April and May 1906, numbered 730 and 731 and classified 4CC, denoting four-cylinder compound (Plate 8.21 and Fig 8.8). These locomotives

Plate 8.20
Class S/07. The most conspicuous difference visible in this last batch of Class S is the adoption of narrower driving-wheel splashers, resulting in the long coupling-rod splasher.

the outer end and a gold line; buffer heads and shanks were black. The engine number was shown at the right-hand side of the drawhook in 6in gold block letters, shaded blue, with 'No' on the left-hand side in similar characters.

Although Class S did not fulfil expectations as express passenger locomotives they were a useful type for excursions and fast goods trains. Ten more were built in 1906, and another 20, to a modified design (Class S/07), in 1908-9 (Plate 8.20).

displayed other features which were new to the NER: both had Belpaire fireboxes while, though No 730 had conventional inside Stephenson link motion, No 731 was fitted with a form of inside Walschaert's valve gear. Unlike the Class V Atlantics, the drive was onto the leading coupled axle, an arrangement later continued on the Class Z 4-4-2s and other three-cylinder Raven locomotives. The 4CCs also had narrow splashers over the coupled wheels, a feature later used on the Worsdell S/07 4-6-0s of 1908-9, the Class V/09 Atlantics, the Raven S2 4-6-0s and other Raven designs.

W M Smith died in October 1906 but in December 1907 Wilson Worsdell was authorised to build ten further compound Atlantics like No 731. However, they were not constructed and it has been said that this was due to the executors of Smith's estate insisting on the

Plate 8.21
Class 4CC. No 731. This originally had the high boiler pressure, for that period, of 225psi, later reduced. The right-hand side of the smokebox differs from that seen here in having a large reducing valve projecting from it.

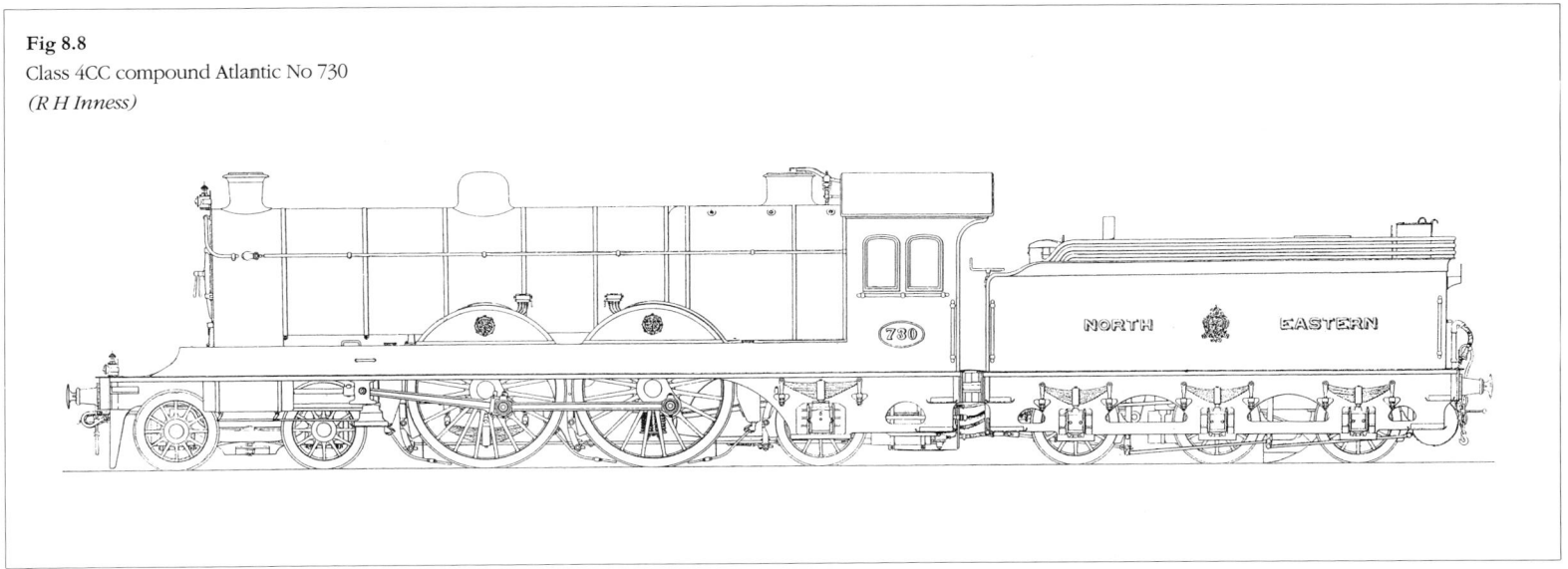

Fig 8.8
Class 4CC compound Atlantic No 730
(R H Inness)

Plate 8.22
Class T. No 2119 in its original green livery. The first ten, including No 2119, came with steam sanding on the driven wheels only, from the large sandbox seen here. That was soon supplemented by leading and trailing sandboxes, the former integral with the front splasher and the latter concealed in the cab - an arrangement depicted in Colour Plate 13. The steam sanding was later changed to gravity feed. The tender seen here originated with Class S as a slightly shortened version of a standard tender. These were used with the first five locomotives (Nos 2116-2120); later ones got the standard tender, both had a water scoop.

payment of royalties. A more likely reason may be an adverse report on Class 4CC by George Heppell, W M Smith's successor as Chief Draughtsman. Heppell, who had a personal dislike for Smith, drew attention to the initial cost of the 4CCs, which had averaged almost £800 more than that of the two-cylinder Class V 4-4-2s built two years earlier, and the maintenance costs of the compounds.[4]

Mineral Locomotives: Classes T and T1

In August 1901 the NER introduced eight-coupled locomotives, as part of the reform of the operating methods for goods and mineral traffic. These 0-8-0s had 4ft 7¼in diameter wheels, 20in by 26in cylinders and 4ft 9in diameter boilers. Ten were built in 1901 with piston valves (Class T) and ten with slide valves (Class T1), followed by 30 more Class T in 1902-04 (Plate 8.22). They all had brass-topped chimneys, brass collars at the front of the boilers and tall brass safety valve covers over 4in diameter safety valves. All but two came out in green livery, with the small heraldic device displayed on the large sandboxes above the driving pair of wheels. The frames were painted green above the running plates (from the leading splashers to the large sandboxes) with a black border and fine white line on the lower edge. The tenders were lettered 'N. E. R.'

Black Livery

In the spring of 1903 an experimental livery was tried on three locomotives. Class T No 1694 and Gateshead Works shunter Class H 0-4-0T No 898 had blue wheels and dark blue panels separated from

---

[4] Further details of W M Smith, and the spread of compounding to the Midland Railway are provided by Philip Atkins in *The Smith Connection*, published in *Midland Record* No 10, 1998, Wild Swan Publishing.

Plate 8.23
Class H. Gateshead Works shunter No 898. It bears what appears to be the 1903 works livery of blue side panels on a black background, with blue wheels; the cab front may also have borne a blue panel. It has received new wheels, the originals having been 'cushion' wheels with much thicker tyres. The chimney in the background belongs to Newcastle & Gateshead Water Company's Askew Road pumping station.

Plate 8.24
Class V. The first of the class, No 532, with its class number in small lettering below the drawhook. Originally it had a flatter arc to the cab roof, but this was soon raised to improve the driver's visibility.

the black border by a pale blue line. Class T No 1696 was painted all black with vermilion lining instead of panels. One of the Darlington Works Class H shunters was also painted in blue livery with the addition of the small armorial device on the side tanks. The treatment of the two Works' shunters was consistent with the adoption of a blue livery for Locomotive Department wagons used in the workshops. The livery of No 1696 evidently satisfied the directors, and, on 4 June 1904, an instruction was issued to the paintshop foreman at Gateshead Works stating that "All goods and mineral engines to be painted black in future, using own discretion as to amount of lining." The first engine to be painted black, following that instruction, was Class P 0-6-0 No 1934. The polished brass of the safety valve cover and the collar at the front of the boiler continued to relieve the black livery. The heraldic device was retained on the sandboxes of Classes T and T1, including the 40 later additions to Class T1 in 1907-08 and 1911; brass-topped chimneys were not fitted after 1908.

As previously mentioned, the 1892 version of the NER heraldic device was displayed on the splashers of some classes of passenger

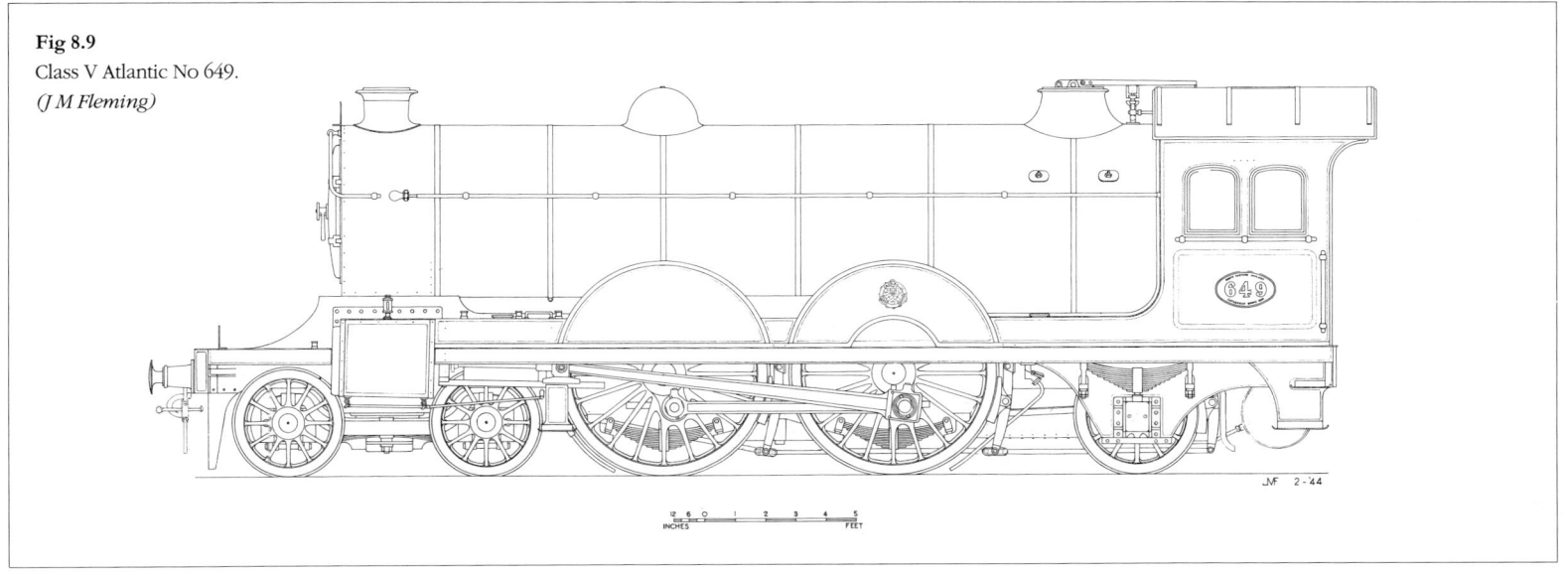

Fig 8.9
Class V Atlantic No 649.
(J M Fleming)

locomotive in conjunction with the larger 1899 style on the tender. However, there was sufficient space for the 1899 device within the black panel on the driving wheel splashers of the Class I and J 'singles' (eg seen on Class I Nos 1327/9, 1528/31 and Class J No 1518). The 1899 device seems to have been used in this way for only a few years before a return was made to the 1892 version. Prior to about 1900 the black panel had been omitted altogether from the splashers of some unrebuilt Class I compounds (eg Nos 1329 and 1530).

Further Wilson Worsdell Developments up to 1910

A programme of rebuilding the two-cylinder compounds, as simple engines with piston valves and Stephenson link motion, was started in 1900. This was soon extended to include the Class F1 and G1 simples, the latter being converted from 2-4-0s to 4-4-0s at the same time. The new cylinders were 18in by 24in, with the valve chests below them. The compound goods engines were also rebuilt as simples, many with slide valves and Joy valve gear, whilst others were given piston valves and link motion.

Another large increase in the size of boilers came about in 1903 with the introduction of the Class V 4-4-2s (Plate 8.24 and Fig 8.9). Wilson Worsdell had been to the USA in October 1901, with other senior officers of the NER, to study American methods of railway operation, and it is probable that this visit may have influenced Worsdell to adopt an Atlantic-type locomotive in preference to developing his 4-6-0 designs. Class V had 6ft 10in diameter coupled wheels, 20in by 28in outside cylinders and a 5ft 6in diameter boiler, the barrel of which was the same length as that of Class S1. The firebox was 9ft long, with the backplate flanged outwards at the edges so that all the riveting could be done from the outside; the roof of the inner firebox was stayed directly to the outer shell. Two pairs of Ramsbottom safety valve columns were mounted on the firebox and enclosed within a handsome polished brass cover.

Ten Class V locomotives were built in 1903-04, the last five having steam-operated reversing gear. All had front buffer beams with 4in thick wood packing between steel plates; the thickness of the wood was tapered down at each end to give a certain amount of elasticity.

The livery of Class V was similar to that of Classes S and S1, the boiler cleading belts and the panels on the cab and tender being edged with white and gold lines; the 1892 heraldic device was displayed on the driving wheel splashers. In September 1901 engines began to have their class letter or number painted inside the cab; from February 1904 this information was shown below the drawhook on the front buffer beam – and on the rear buffer beam of the tank engines – in white letters 1½in high (gold letters for Classes S, S1, V etc). In addition, from September 1905, the traffic classification, ie the haulage capacity, was painted below the number plates. Passenger engine classifications were T1 to T10 (for tender engines) and K1 to K3 (for tanks), and for freight engines T21 to T26 and K21 to K23, respectively. This practice was discontinued after June 1910.

As previously mentioned, a black livery was adopted in June 1904 for goods and mineral engines. An additional instruction was issued on 29 June: "Inside of all engine frames to be painted vermilion until further notice" (possibly this was to confirm that the existing use of vermilion, for engines in green livery, would also extend to those in black). A later order, dated 20 January 1905, applied only to goods, mineral and shunting engines: "In future the handrails, smokebox hinges, knobs, side rods etc will not be polished". This instruction was not strictly observed however, many smokebox door hinges remaining polished instead of painted!

A much enlarged version of the Class P1 0-6-0s was introduced in 1904. This was Class P2, which had the same size of wheels and cylinders as the later engines of Class P1 built in 1902, but with a boiler 5ft 6in in diameter. The engine frames were lengthened by 11in at the rear to accommodate a firebox 7ft long, its construction being similar to those of Class V – including the sloping grate.

Class P2 numbered 50 engines, built between June 1904 and December 1905. These were followed by the very similar Class P3 0-6-0s, which had a deeper firebox and some minor modifications (Plate 8.25). Darlington Works built 30 of the latter class between April 1906 and December 1908, and 50 more were supplied by three different contractors between May 1908 and September 1909.

Plate 8.25
Class P3. No 1014 in grey. The tender has breather pipes, with bonnets on top, for water pick-up but there is no sign of a scoop.

Plate 8.26
Class R1 No 1238, seen in grey. It had been Worsdell practice to form the smokebox with a larger external diameter than the clad boiler and with an integral, shouldered base. Class R1 introduced a 'drum-head' or fully circular smokebox, flush with the boiler cladding and resting on a separate saddle. This was used also on Class V/09 and later on all Vincent Raven's tender locomotives.

The Class P2 and P3 0-6-0s all came out in black livery, with vermilion lining; buffer beams were vermilion with a black border but no white lining. Tenders bore the initials 'N. E. R.' in gold block letters shaded red. The twin safety valves had large brass covers, like those of Class V, and a brass collar encircled the front end of the boiler. There was no brass top on the chimneys of Class P2 and P3 although the 20 black-liveried Class T1 0-8-0s built in 1907-08 came out with brass-topped chimneys and the NER 1892 heraldic device on their large sandboxes. (The final 20 Class T1 engines, built in 1911, had plain chimneys.)

Even after the adoption of the black livery, goods engines were painted grey for official photographs and usually fully lined out as though in passenger livery. This practice, which has confused not a few artists and model makers, began with Class P2 No 132 in June 1904 and still persisted as late as November 1921 when photographs were taken of the new superheated P3 No 2338.

Fig 8.10
Class R1, No 1238, *(J M Fleming)*

Plate 8.27 (left)
Class V/09. No 696 was completed in 1910 and the coupling-rod splashers are given prominence by the adoption of narrower driving-wheel splashers than in the original Atlantics.

Plate 8.28 (below)
Class 901. In 1907 Worsdell experimentally rebuilt this Fletcher Class 901 2-4-0 No 933 as a handsome 4-4-0, beside which the old tender looked very incongruous. The locomotive only survived until 1914.

The success of the Class R 4-4-0s led to the construction of a larger version, Class R1, whose ten locomotives appeared in 1908-9 as Nos 1237 to 1246 (Plate 8.26 and Fig 8.10). The increase in size was largely due to the adoption of a 5ft 6in diameter boiler developed from that used by the Class V Atlantics. It had a grate area of 27 square feet and, originally, a working pressure of 225psi, soon to be reduced to 180psi. Class R1's use on main-line expresses was, however, short-lived as the work was taken over by Atlantics, particularly the Class Z which came out in 1911.

Ten further Atlantics, this time built at Darlington Works, were added to the locomotive stock in 1910. They were classified V/09, since the order was placed in 1909 and their design exhibited some detail differences from the Class V of 1903-4. The cylinder diameter was 19½in, in place of 20in, and the splashers over the coupled wheels were of the narrow variety pioneered on the Smith compound 4-4-2s in 1906 (Plate 8.27).

From December 1907 to April 1908 Gateshead Works turned out ten inside-cylinder 4-6-0 tank locomotives of Class W with the intention

Plate 8.29
Class W. No 695, one of the second batch of five, built with a normal smokebox and provided with two coal rails; the first five were built without these rails but altered to conform. The Westinghouse pump was on the right hand side of the smokebox. This view is in grey, but the lining corresponds to that adopted throughout their working existence, even after rebuilding as 4-6-2Ts.

Plate 8.30
Class X. No 1352, seen in grey. Their working livery was black, lined in vermillion as depicted in Colour Plate 8. *(Ken Hoole collection)*

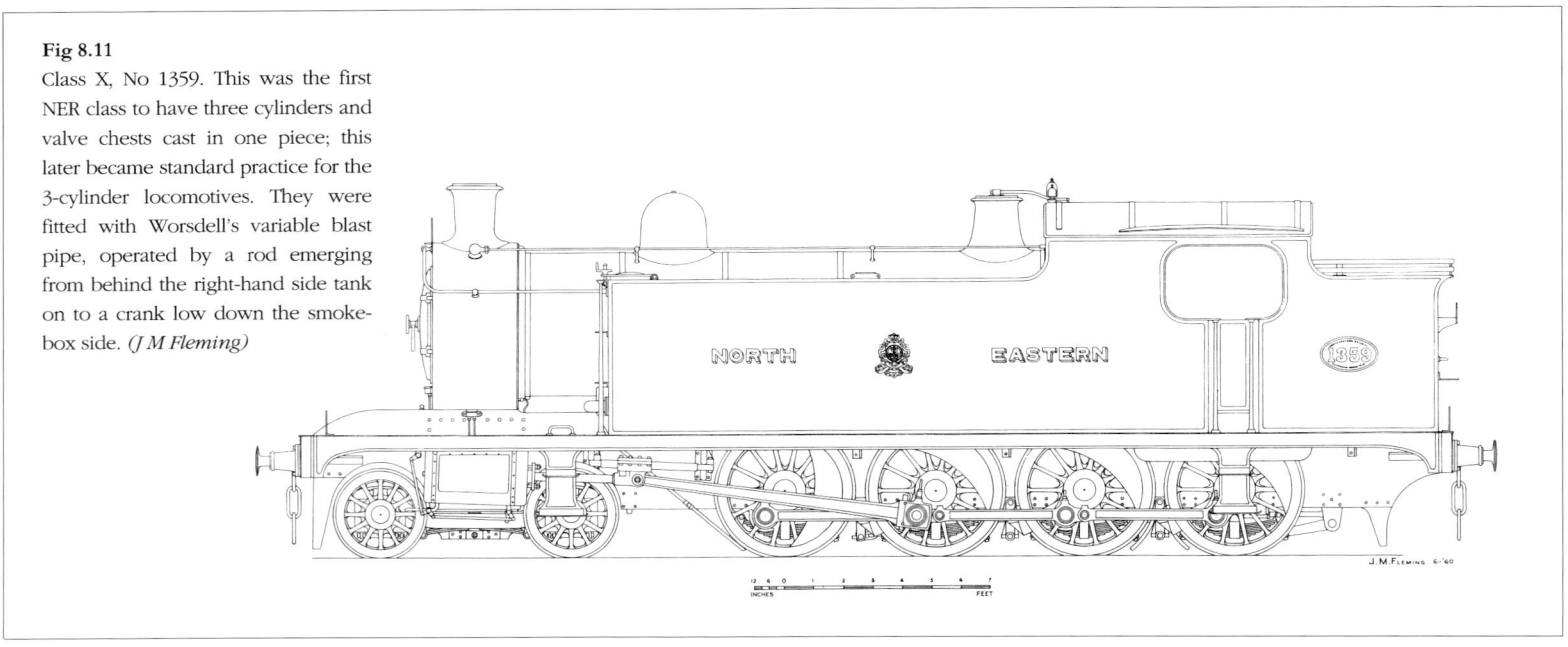

Fig 8.11
Class X, No 1359. This was the first NER class to have three cylinders and valve chests cast in one piece; this later became standard practice for the 3-cylinder locomotives. They were fitted with Worsdell's variable blast pipe, operated by a rod emerging from behind the right-hand side tank on to a crank low down the smokebox side. (*J M Fleming*)

that they would take over much of the work then being done by the Class O 0-4-4 tanks on two difficult routes: the coast line between Saltburn and Scarborough and the line between Whitby and Malton (Plate 8.29). They were numbered 686 to 695. It would seem that the 4-6-0 wheel arrangement was chosen in order to restrict the length of the wheelbase but this only allowed for a small rear coal bunker, which was to prove inadequate. The first five locomotives had extended smokeboxes, giving a very ungainly appearance, but the remainder appeared with normal smokeboxes. In 1914 No 688 was rebuilt as a 4-6-2 tank, with a larger bunker giving increased coal capacity and a well tank at the bottom providing a further 393 gallons of water. The other nine members of the class were similarly rebuilt in 1915.

The last locomotives to be built at Gateshead Works were turned out between September 1909 and April 1910; these were the three-cylinder 4-8-0 tanks of Class X, numbered 1350 to 1359 (Plate 8.30 and Fig 8.11). New locomotive construction was then transferred to the much-expanded Darlington Works and Gateshead was left with repair work.

With 18in by 26in cylinders and 4ft 7¼in diameter coupled wheels, the Class X locomotives weighed over 84 tons each, had a tractive effort of 34,080 pounds and, until the arrival of the Class T3 0-8-0s in 1919, were the most powerful locomotives on the NER. Their construction was brought about by the need for more powerful shunting locomotives to handle trains of shipment coal at Tyne Dock, Newport (Middlesbrough), Gascoigne Wood (near Selby), and Hull. Much of their work consisted of sorting wagons by propelling train-loads at low speed over the hump to allow the wagons to run into the appropriate sidings.

The NER was a pioneer of electric traction, and in 1904 two Bo-Bo electric locomotives, operating at 600V dc were built for working the steep Newcastle Quayside branch (Plate 8.31). They had a central cab

Plate 8.31
Class Electric 1. No 2, one of the two Quayside Branch locomotives with its original bow collector, used for current pick-up outside the tunnels and later replaced by a pantograph, as well as the third-rail pick-up shoes. 'CLASS ELECTRIC 1' originally appeared on the bufferbeam below the number but cannot be made out. This locomotive was withdrawn in 1964 and scrapped, but its sibling, No 1, now resides in the National Railway Museum.

Plate 8.32

Londonderry Railway. No 21, built at Seaham in 1895, became NER No 1712 but was sold to the Isle of Wight Central Railway in 1909, becoming their No 2. Resold to Armstrong Whitworth in August 1917 for use at their Elswick Works on Tyneside, it was scrapped about 1921. It had 5ft 4 in coupled wheels. The side tanks are adorned with the Marquis of Londonderry's coronet within a garter.

and were equipped to collect current either from third rail or overhead wire. These locomotives came out in green livery which they retained for many years. The sides and top of the end bonnets, and the sides of the cab, were each separately panelled with black bands edged with white lines; the bonnets, but not the cab, also had black borders with white lines, the sides being lettered 'NORTH EASTERN'. The wooden doors of the cab were green and bore the '1892' heraldic device. The soles were green, with black borders and white lines. Buffer beams were red, with a narrow black border and 'No' and the number on opposite sides of the drawhook; it seems that these letters and numerals were originally black with light (possibly blue) shading, and 'CLASS ELECTRIC 1' was in 1½in black letters. Number plates were not fitted. These locomotives were repainted black eventually, but retained the 'NORTH EASTERN' lettering.

Londonderry Railway

The Londonderry Railway (LR) was taken over by the NER in October 1900 and later became part of a new route between Sunderland and West Hartlepool. Three fairly new locomotives which had been built at the LR Works at Seaham were taken into NER capital stock. Ten older locomotives were given duplicate stock numbers 2267-76 and were scrapped or sold within a few years. From quite an early date the LR locomotives had large cabs with a window in each side. (Plate 8.32)

The LR locomotive livery was emerald green, with black borders and panels, and orange yellow lining. In the 1860s a cast brass representation of the Londonderry family's coronet was affixed to the sides of the tenders. In later years a painting, or transfer, of the Londonderry coat-of-arms within a garter appeared on the tenders or the side tanks of the passenger engines; the initials 'L R', with or without the coat-of-arms, was also to be seen, and sometimes there was no indication of ownership at all!

Chapter 9

Locomotive Progress 1910 – 1922
Vincent Litchfield Raven, Chief Mechanical Engineer

Wilson Worsdell retired at the end of May 1910 and was succeeded by Vincent L Raven, who had been Assistant Chief Mechanical Engineer since 1893. Raven started as a pupil of Edward Fletcher and, except for his wartime service, spent his entire career in the NER Locomotive Department. In September 1915 he was seconded to the Royal Ordnance Factory at Woolwich, as Chief Superintendent, and in May 1917 became Deputy Controller of Armament Production at the Admiralty. He was knighted in February 1917 and created K.B.E. in August 1917. Sir Vincent Raven returned to the NER soon after the Armistice and was the company's last CME.

Raven favoured the use of three cylinders, which gave a more even torque, a steadier blast on the fire and better balancing of the rotating parts. Short connecting rods, each driving the leading pair of coupled wheels, were a feature of all but one of his 3-cylinder designs, the exception being the 0-8-0s on which the drive was on the second pair. Raven was not only a very progressive locomotive engineer but also an enthusiastic advocate of railway electrification.

Developments up to 1914

The Class Y 4-6-2T was the first new class of locomotives to appear after Raven's appointment as CME. Although ordered by his predecessor, these 3-cylinder mineral engines showed many signs of Raven's influence in their design (Plate 9.1 and Fig 9.1).[1]

The first new design initiated by Raven was his highly successful 3-cylinder express passenger 'Atlantic' 4-4-2 (Plates 9.2 & 9.3). Ten of these were built in 1911 with saturated boilers (Class Z) and ten with superheated boilers (Class Z1); during 1914-15 the saturated boilers were provided with superheaters, and all the engines were classified 'Z' from June 1914. Further additions were made to Class Z between 1914 and 1918, those built from December 1914 having 'self

---

[1] They were designed to haul 1,000 tons on the level at 20mph.

Plate 9.1
Class Y. No 1175, built at Darlington in 1910, seen in its original black livery, lined red, with the large armorial device. A blower control rod can be seen running along the left-hand side of the boiler; on other classes this was normally concealed within the boiler handrail.

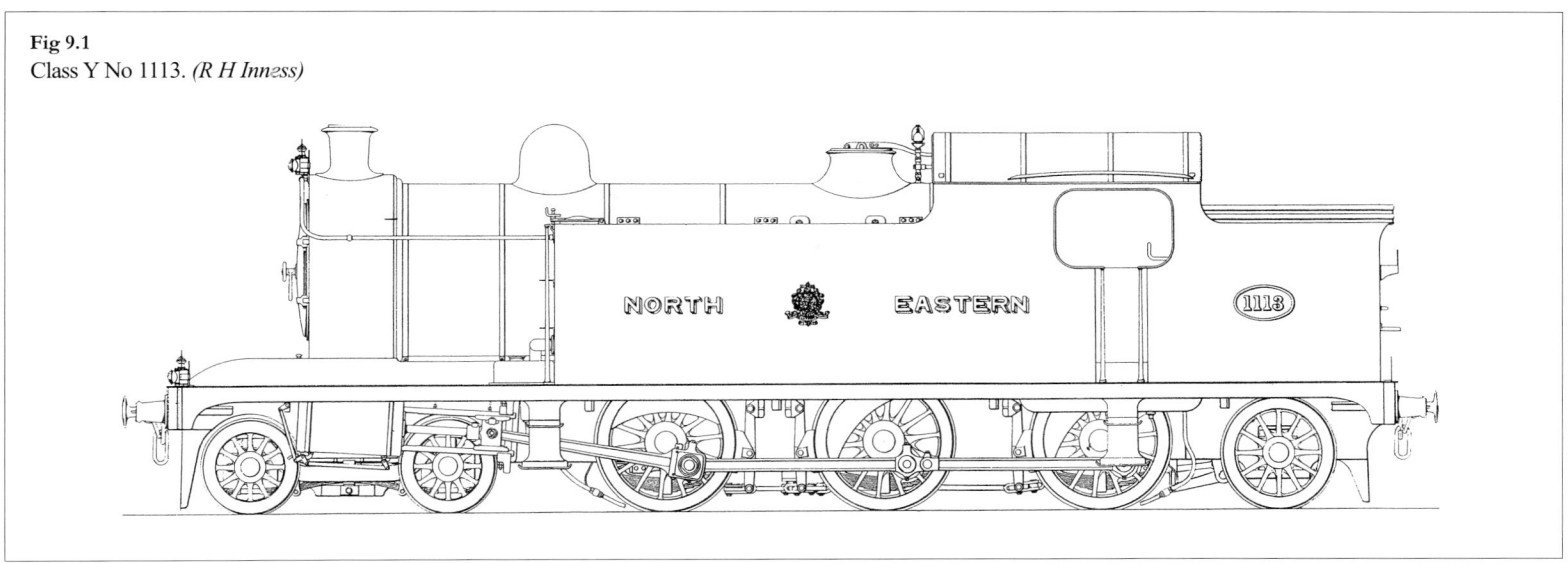

Fig 9.1
Class Y No 1113. *(R H Inness)*

Plate 9.2
Class Z. No 717, built by the North British Locomotive Company in Glasgow in 1911, seen in the maker's official photograph, painted in grey.

Plate 9.3
Class Z. Cab and front views of No 717 taken by the North British Locomotive Company, whose photographer has posed targets above the buffers.

Plate 9.4
Class S2. Its normal appearance, with No 797 in the black livery adopted for the class in May 1912. It was the first S2 to be built with a superheater, and a control rod to a steam circulating valve can be seen running along the boiler side from the cab. Between this and the handrail runs a sheathed cable connecting a gauge in the cab to a pyrometer, monitoring the temperature of the superheated steam.

Plate 9.5
Class T2. The first of its class, No 1247 photographed in grey in February 1913 in the paint shop yard at Darlington. Its working livery was black.

trimming' tenders in which the inward sloping sides and back of the coal space formed a chute to deliver the coal towards the fireman.

Raven's next two designs were enlargements of existing types, each having two cylinders. Class S2, introduced in 1911 for passenger and fast goods trains, was basically similar to the Class S/07 4-6-0s of 1908-09, but had a 5ft 6in diameter boiler with the smokebox mounted on a saddle (Plate 9.4). The first seven engines used saturated steam, but the remaining thirteen had superheated boilers. Class T2, introduced in 1913, was developed from the Class T piston valve 0-8-0s of 1901-04 and incorporated a 5ft 6in diameter superheated boiler of the type fitted to Class S2 (Plate 9.5).

Fig 9.2
Class S2 No 825 with Stumpf 'Uniflow' cylinders.
(J M Fleming)

Plate 9.6
Class S2. No 825, somewhat deformed to accommodate 'Uniflow' cylinders. It had Walschaert's valve gear.

The final engine of Class S2 – No 825, built in 1913 – had Stumpf 'Uniflow' cylinders with the valves placed above them and actuated by outside Walschaerts gear (Fig 9.2). Live steam was admitted at the ends of the cylinders and exhausted at the centre through a series of radial holes that were uncovered by the pistons at each end of their stroke. The cylinders and valve chests were twice the normal length and had to be raised up to clear the bogie wheels; that part of the running plate above the huge cylinder casting and the valve gear was 7½ft above rail level, giving the engine a rather ungainly appearance (Plate 9.6). The final engine of Class Z – No 2212, built in 1918 – also had 'Uniflow' cylinders with the valves above them, but with inside valve gear, the valves being actuated by separate sets of Stephenson valve gear and rocking shafts; a non-standard bogie with a long wheelbase and small wheels enabled the cylinder casting to be accommodated at near normal height (Plate 9.7 and Fig 9.3).

Raven's passenger tank locomotives were the Class D 4-4-4Ts with three cylinders and superheated boilers; these first appeared in 1913 and were originally intended to replace some of the older tender locomotives on branch lines (Plate 9.8). Raven also built 20 more Class E1 0-6-0T locomotives at Darlington in 1914 with enlarged bunkers and tank vent pipes. The Ramsbottom safety valves and bell-type whistle on the earlier Class E1 locomotives were replaced by Ross pop safety valves and an organ-type whistle in the new version

Plate 9.7
Class Z. The final locomotive, No 2212, equipped with Stumpf 'Uniflow' cylinders, hence the extended bogie wheelbase and the projection above the running plate.

(Fig 8.7b). Further locomotives of Class E1 were built for the NER and LNER up to 1925, while a final 28 were built by British Railways at Darlington in 1949-51.

Detail Changes

The Class Z and Z1 4-4-2s built in 1911 were the last new NER locomotives to be fitted with brass-capped chimneys. Together with the Class T1 0-8-0s built in 1911, these 4-4-2s were also the last locomotives to be provided with Ramsbottom safety valves encased within an ornamental brass cover, although new boilers with Ramsbottom valves were built as replacements for a few more years. Subsequent locomotive designs had Ross 'pop' safety valves with only a low cover over the base.

In March 1911 Gateshead Works received an instruction to use sheet iron instead of brass for the beading around the firebox backplate of all classes of locomotives. Two months later, the order was given to fit chimneys without brass caps to Class S 4-6-0s when replacements were required. The latter instruction applied subsequently to all classes except the four 2-2-4Ts that were used for hauling the various Officers' Saloons. It is possible that 1911 may also have been the year in which the simulated wood finish on the interior of the cabs was superseded by a plain colour, although the actual date when this change took place is uncertain. Latterly, Darlington Works used a medium brownish-red colour, whereas Gateshead favoured Saxony green with a black border and white line – at least for its passenger locomotives.

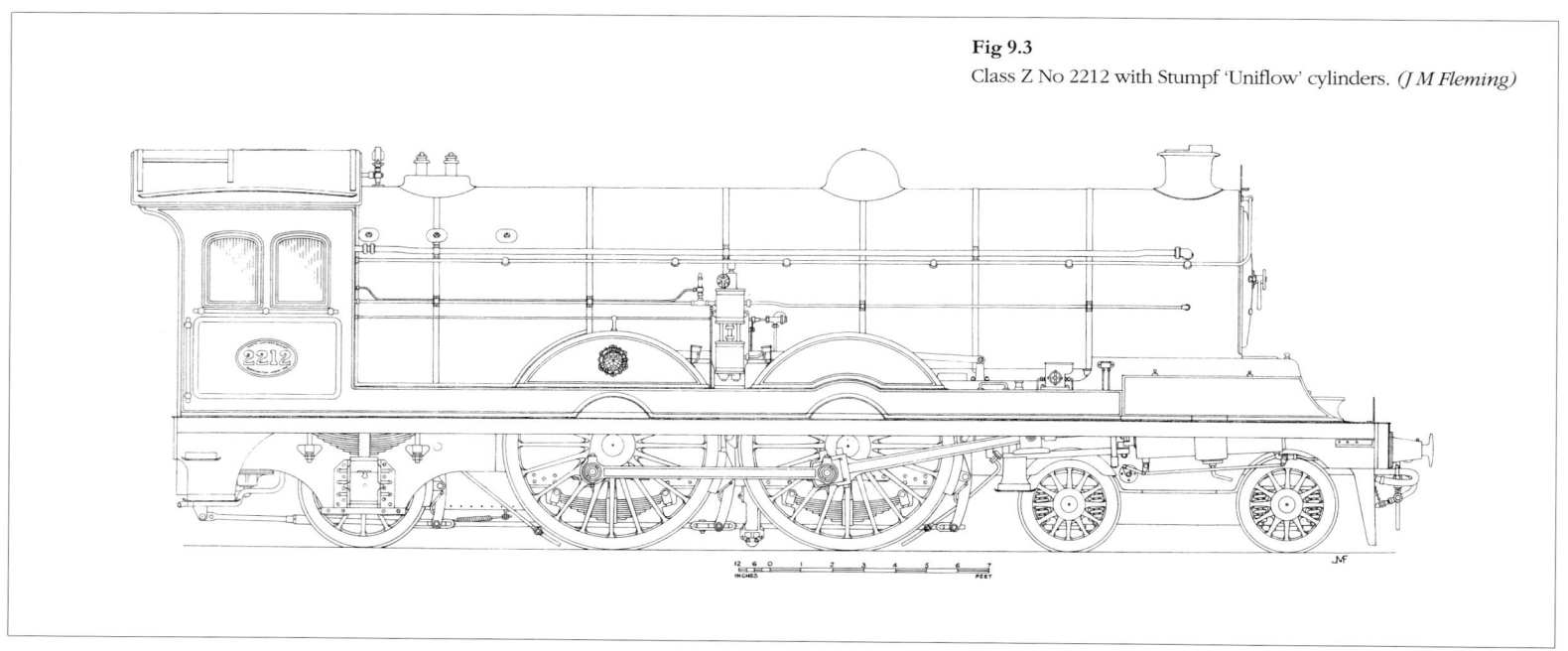

Fig 9.3
Class Z No 2212 with Stumpf 'Uniflow' cylinders. *(J M Fleming)*

Plate 9.8
Class D. No 2147, seen at Neville Hill shed, Leeds. A bonneted breather pipe is visible on top of the side tank; there was one each side and a third for the bunker tank. The Westinghouse pump was mounted on the right-hand side of the smokebox, with the steam reverse cylinders between it and the tank

Plate 9.9
Class P3. No 1172 bearing the smokebox front numbers adopted on a trial basis in 1912.

Classes Z and Z1, and the first seven locomotives of Class S2, were painted green and had the white and gold lining appropriate to principal passenger locomotives. However, in May 1912, a 'special' black livery was introduced for Class S2, the usual $1/8$in wide vermilion lining being supplemented by $3/8$in gold lines on the cab sides, splashers, boiler cleading belts, cylinder cleading and tenders; the brass splasher beadings were painted black. The tender sides were lettered 'NORTH EASTERN', with the large heraldic device in the centre, whilst the small device – without the usual border – was on the driving wheel splashers. (The official photograph of No 797 and No 825 (Stumpf) in 'photographer's grey' paint do not show the true lining.) From 1912 the 'special' black livery was also applied to the Class S and S1 4-6-0s, which by that date were increasingly being used for fast goods trains, but in some cases their brass splasher beadings remained polished.

The adoption of the Train Control system by the NER led to a trial painting of the engine number on Class P3 No 1172 in 1912 to assist in identification. Large white numerals were painted on the smokebox door and the back of the tender, lamps being provided to illuminate these at night. (Plate 9.9) It was proposed to treat all NER locomotives in this manner, so that signalmen, where necessary, could easily distinguish the number, but this is thought to have been the only one so fitted.

Developments from 1915 to 1922

1915 was notable for the introduction of electric traction for the intensive mineral traffic from Shildon to Newport. Raven was responsible for the mechanical design of the ten electric locomotives, which were built by the NER at Darlington Works, with electrical equipment supplied by Siemens. The system was 1500V dc overhead. The original livery of the electric locomotives was black with vermilion lining; they were lettered 'NORTH EASTERN' on the sides and had brass numberplates on the end bonnets. (Plate 9.10) The cab doors were of varnished wood, with the small heraldic device on the panel below the window.

After March 1917, all goods and mineral locomotives – including the 4-6-0s and at least one of the Class 4CC 4-4-2s – were painted black, without lining or heraldry, and most of the large brass numberplates were replaced by smaller ones. The locomotive numbers were painted on the tenders or side tanks in figures 12in high, with the initials 'N. E.' in 6in letters on either side. (Fig 9.4) The figures and letters were yellow, with a ⅛in white line and 1in red shading on the right-hand side and below. The Shildon-Newport electric locomotives retained the 'NORTH EASTERN' lettering and the large brass numberplates, with the addition of 12in high numbers painted on the cab doors in place of the heraldry.

Shunting locomotives retained their large brass numberplates and the initials 'N. E. R.', but were without vermilion lining. Certain shunters engaged on yard-to-yard interchange workings came under the traffic control system and had 12in painted numerals in addition to their large brass numberplates.

Plate 9.10
Electric Freight Locomotive. No 5, built for the Newport-Shildon working, in store at Gosforth Car Sheds prior to being scrapped in 1951. Throughout LNER ownership it had retained its NER numberplate and livery, while British Railways had simply repainted a token area to accommodate its new number 26504.

Preliminary design work was in hand during 1917 for three proposed new classes of locomotives, comprising two types of 4-6-0 – for express passenger and mixed traffic, respectively – and an 0-8-0 for mineral trains. Each of these classes would have had three cylinders 18in diameter by 26in stroke and a standard design of boiler and firebox, based on or similar to that of Class Z. The piston heads and rods, crossheads, connecting rods, valve gear, axles and many other parts were to be common to all three classes. The diagrams (Fig 9.5) illustrating these proposals show that all would have had straight

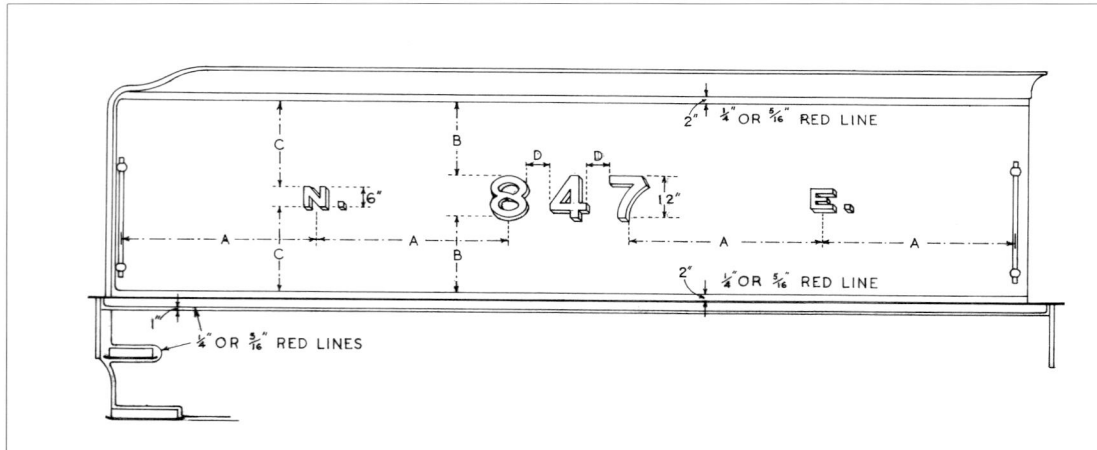

Fig 9.4
Locomotive numbers painted on tenders after March 1917. The letters denote dimensions of equal size; for any particular tender the actual measurements have to be scaled from its length and height. (*J M Fleming*)

running plates which, in the case of the two 4-6-0s, would have been surmounted by a raised platform and narrow splashers, as on the Class Z 4-4-2s.

The Class S3 mixed traffic 4-6-0s and the Class T3 mineral locomotives, both of which were introduced in 1919, were developed from the 1917 proposals; the boiler dimensions and the respective wheel diameters and wheelbases were unchanged, but the diameter of the cylinders was increased to 18½in and there was an increase of 3in in the length of the connecting rods. The overall length of both classes was also increased, and they had raised running plates. The proposed express passenger 4-6-0 never materialised. (Plates 9.11 & 9.12) Their livery was black, lined in vermilion.

Other post-war construction comprised additions to existing classes, including a modified version of the Class P3 0-6-0 with piston valves and superheated boiler, and more rebuilds from Class BTP to Class 290 0-6-0T.

Provisional approval was given, late in 1919, for the electrification of the East Coast Main Line between York and Newcastle. Raven went ahead and designed an experimental electric express locomotive – No 13 – which was completed at Darlington Works in May 1922, with electrical equipment supplied by Metropolitan-Vickers. Unfortunately, economic uncertainties caused the abandonment of the electrification scheme. No 13 ran trials in grey paint with full lining-out, but does appear to have been painted in the NER Saxony green

Plate 9.11 *(below)*
Class S3. The NER's most successful 4-6-0 class. No 846, completed at Darlington in 1920, is seen in grey, rather than its working livery of black, lined red.

Plate 9.12 *(bottom)*
Class T3. The first of this small class of powerful mineral engines, No 901, emerged from Darlington in 1919 and is seen in grey. After withdrawal in 1962 it was preserved in the National Collection and can currently be found on the North Yorkshire Moors Railway.

4-6-0
PROPOSED STANDARD EXPRESS ENGINE
SUPERHEATED

3 CYLINDERS 18" DIA. 26" STROKE

BOILER BARREL 16'-8" LONG 5'-6" DIA. — BETWEEN TUBEPLATES 16'-2 5/8"
BOILER PRESSURE 180 p.s.i.   FIREBOX 9'-0" LONG
MAXIMUM TRACTIVE EFFORT AT 80% BOILER PRESSURE 22,744 lbs OR 10·15 TONS — ADHESIVE FACTOR 5·1

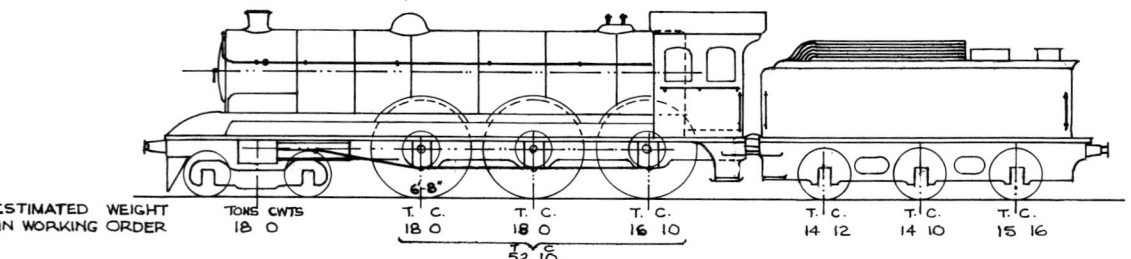

	TONS CWTS	T. C.	T. C.	T. C.		T. C.	T. C.	T. C.
ESTIMATED WEIGHT IN WORKING ORDER	18 0	18 0	18 0	16 10		14 12	14 10	15 16
			T C 52 10					

ESTIMATED TOTAL WEIGHT OF ENGINE IN WORKING ORDER TONS 70 CWTS 10   SELF TRIMMING (STANDARD) WITH WATER PICK-UP.
    "   "   "   " TENDER "   "   " 44 18   CAPACITY OF TANKS 4125 GALLONS
    "   "   "   " ENGINE & TENDER "   " 115 8   "   " COAL BUNKER 5 TONS

4-6-0
PROPOSED STANDARD GOODS ENGINE
SUPERHEATED

3 CYLINDERS 18" DIA. 26" STROKE

BOILER BARREL 16'-8" LONG 5'-6" DIA. — BETWEEN TUBEPLATES 16'-2 5/8"
BOILER PRESSURE 180 p.s.i.   FIREBOX 9'-0" LONG
MAXIMUM TRACTIVE EFFORT AT 80% BOILER PRESSURE 26,758 lbs OR 11·94 TONS — ADHESIVE FACTOR 4·3

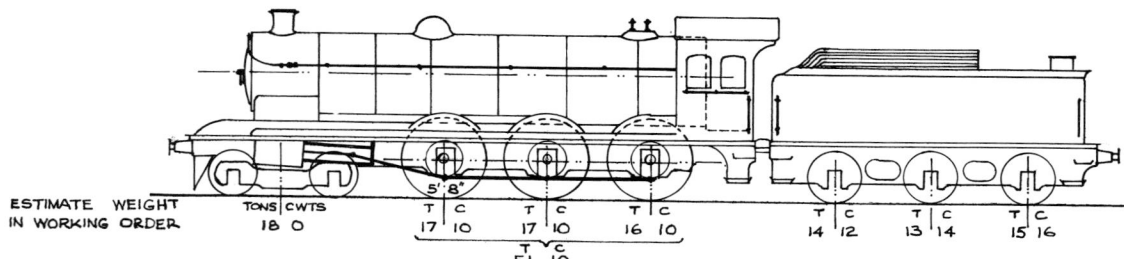

	TONS CWTS	T. C.	T. C.	T. C.		T. C.	T. C.	T. C.
ESTIMATE WEIGHT IN WORKING ORDER	18 0	17 10	17 10	16 10		14 12	13 14	15 16
			T C 51 10					

ESTIMATED TOTAL WEIGHT OF ENGINE IN WORKING ORDER TONS 69 CWTS 10   SELF TRIMMING TENDER (STANDARD)
    "   "   "   " TENDER "   "   " 44 2   CAPACITY OF TANKS 4125 GALLONS
    "   "   "   " ENGINE & TENDER "   " 113 12   "   " COAL BUNKER 5 TONS

0-8-0
PROPOSED STANDARD MINERAL ENGINE
SUPERHEATED

3 CYLINDERS 18" DIA. 26" STROKE

BOILER BARREL 16'-8" LONG 5'-6" DIA. — BETWEEN TUBEPLATES 16'-2 5/8"
BOILER PRESSURE 180 p.s.i.   FIREBOX 9'-0" LONG
MAXIMUM TRACTIVE EFFORT AT 80% BOILER PRESSURE 32,492 lbs OR 14·5 TONS — ADHESIVE FACTOR 4·89.

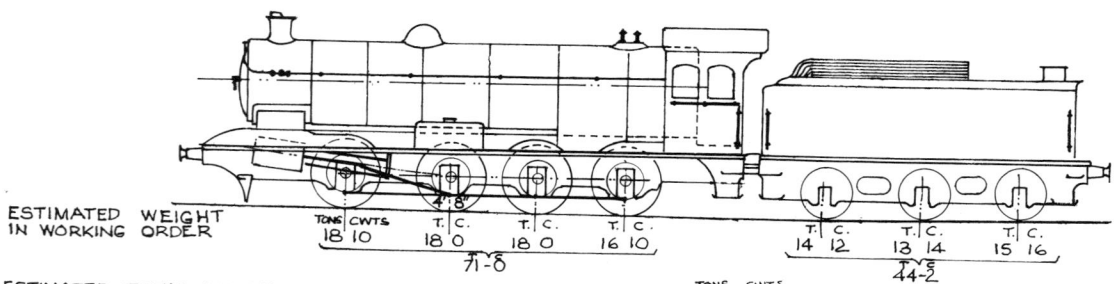

	TONS CWTS	T. C.	T. C.	T. C.		T. C.	T. C.	T. C.
ESTIMATED WEIGHT IN WORKING ORDER	18 10	18 0	18 0	16 10		14 12	13 14	15 16
			71 8			T C 44 2		

ESTIMATED TOTAL WEIGHT OF ENGINE & TENDER IN WORKING ORDER TONS 115 CWTS 2

SELF TRIMMING TENDER (STANDARD)
WATER 4125 GALLONS   COAL 5 TONS

NOT TO SCALE

Plate 9.13
Express electric locomotive No 13. It was designed to haul 450 tons on the level at an average speed of 65mph, with a peak capability of 90mph. The bonnets at each end sloped inwards on plan to cope with the curved platforms of York and Newcastle. Though the main-line electrification was cancelled, the LNER continued to exhibit No 13 and it was left to British Railways to order the destruction of this pioneering locomotive in 1950.

Fig 9.5 *(opposite)*
Three locomotive designs in 1917, sharing standard components. The first, a passenger design, remained unbuilt; the second developed into the Class S3 and the third into Class T3. *(J M Fleming Collection)*

Fig 9.6
Express passenger electric locomotive No 13. *(R H Inness)*

livery, with lemon chrome lining, in the autumn of 1922. (Plate 9.13 and Fig 9.6)

With the end of the independent existence of the NER in sight, and the prospect of main line electrification already fading, Raven brought out his last express passenger steam locomotives – the impressive 4-6-2 'Pacifics' – two of which were completed by December 1922. (Plate 9.14 and Fig 9.7)

A Livery Postscript

Although the independent existence of the NER came to an end on 31 December 1922 when it became part of the London & North Eastern Railway (LNER), it is relevant to mention some early post-grouping liveries based on NER styles. In the first weeks of 1923 a

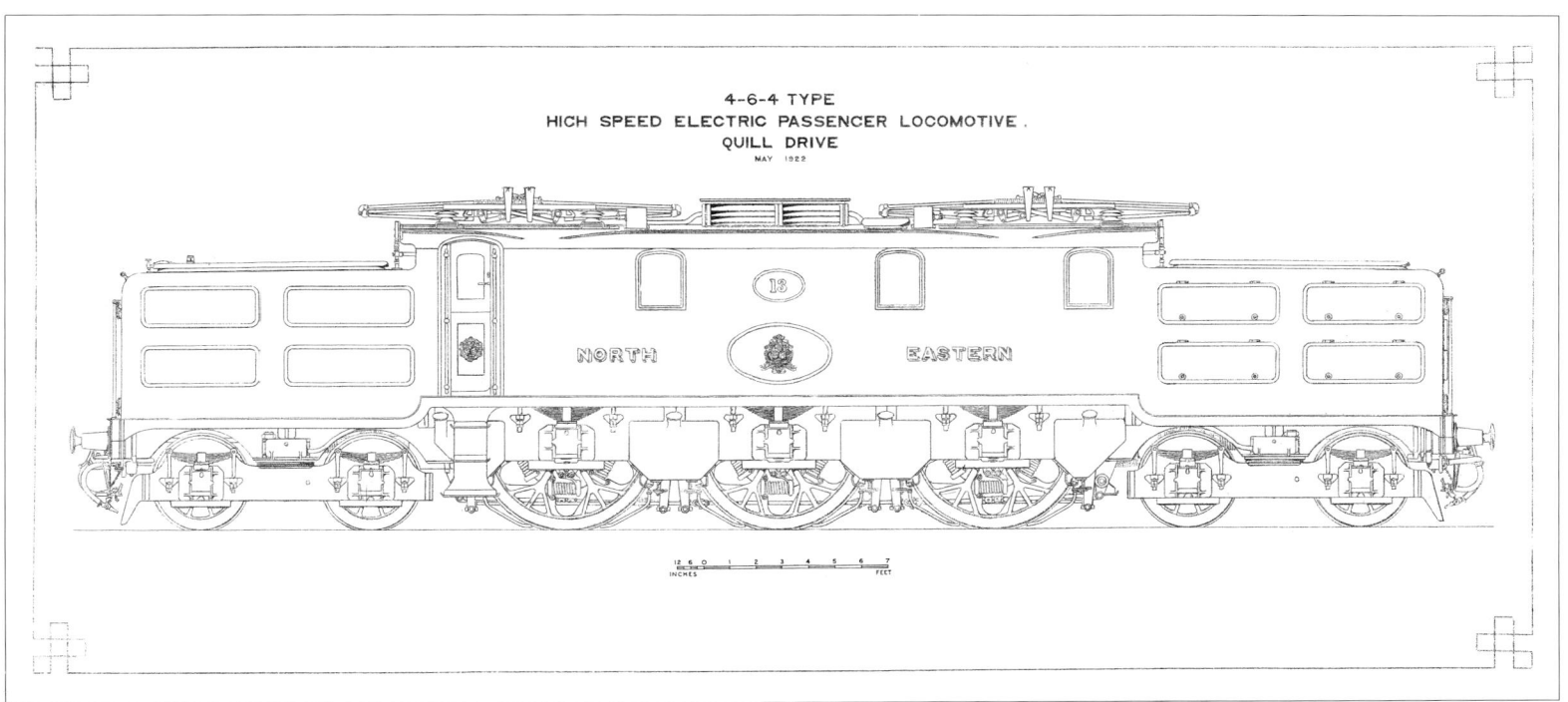

few passenger locomotives were turned out in NER green livery, complete with large brass numberplates. The heraldry was omitted altogether and the tenders bore the initials 'N. E. R.' in 6in letters with the locomotive number in 12in figures below. The letters and figures were lemon gold with red shading.

Eight locomotives, from four of the constituent companies of the LNER, were painted in a variety of different liveries and displayed at York for inspection by the Directors on 31 January 1923. Two of these locomotives were of NER origin: Class Z No 2169 was painted

Plate 9.14
Class 4-6-2. Raven's first Pacific, No 2400, was completed in November 1922. The LNER named it *City of Newcastle* in 1924 and placed a short, curved nameplate above the centre splasher. Two were built by the NER and three more by the LNER but they were edged out by the Gresley Pacifics, and No 2400 was withdrawn in 1937. The buffer beam originally had square ends but is seen with a rounded cut-out to avoid fouling the platforms at Newcastle.

Fig 9.7
Raven Pacific No 2400. A presentation drawing made by R H Inness for the North Eastern Railway Boardroom at York.

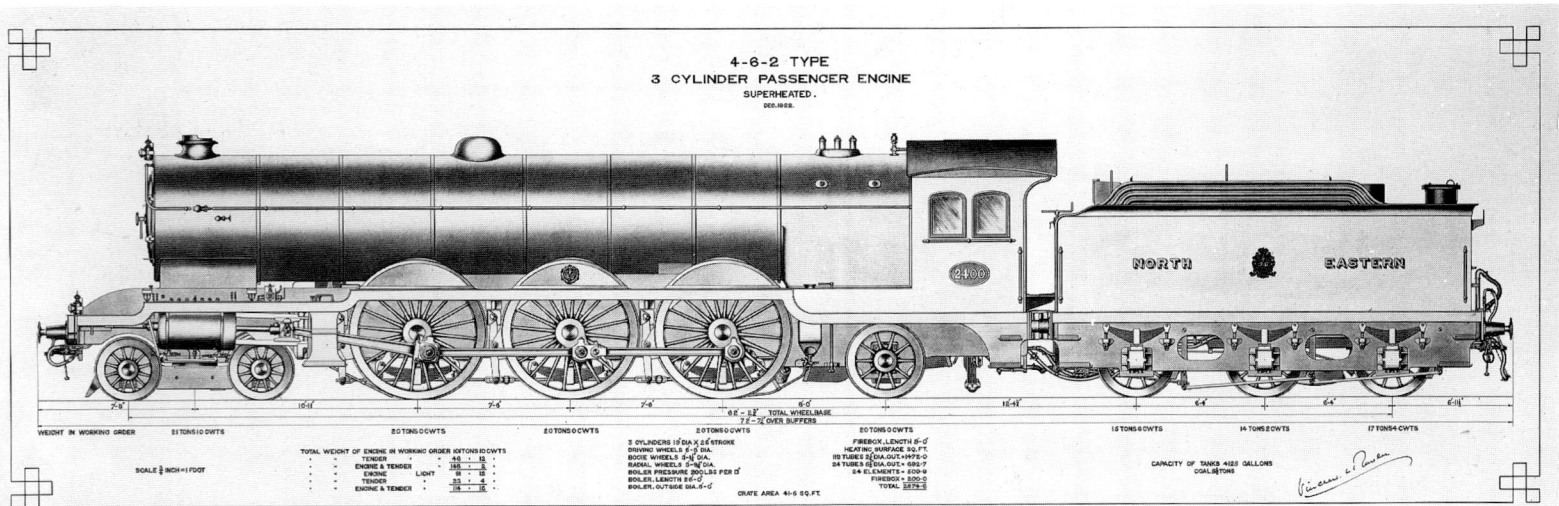

Saxony green, with NER-style panelling and white and pale chrome lining and large brass NER-pattern numberplates; the tender sides carried the number in 12in 'lemon gold' figures, shaded red, with the initials 'L.&N.E.R.' in 6in letters above: Class T2 No 2237 was painted black, with red lining in NER style, and had large brass NER-pattern numberplates with the new company's name; the number and initials on the tender were similar to those on No 2169 except that yellow was substituted for 'lemon gold'.

Another inspection took place at Marylebone Station in London on 22 February 1923, when 13 locomotives were exhibited. The NER representatives again included No 2169, together with Class Z No 2207 in Great Northern Railway (GNR) apple green, with NER-style white lining and small NER-pattern numberplates on the cab; the tender was lettered and numbered like that of No 2169: NER goods locomotives were represented by Class S3 No 2368 in black livery similar to No 2237, but with small NER-pattern numberplates.

The apple green livery of No 2207 was adopted by the LNER for most of its principal passenger locomotives, and the black livery of No 2368 for other classes. The size of the letters 'L.& N.E.R.' was increased to 7½in. The new LNER numberplates were similar to the small NER post-1917 pattern. A few non-standard arrangements were to be seen in 1923, as for example on Class E 0-6-0T No 978, which retained its large NER numberplates in conjunction with 'L.& N.E.R.' in 7½in letters on the side tank.

'L.& N.E.R.' soon gave way to 'L&NER' without 'stops' and, by August 1923, to 'LNER'. Further changes were made at a later date, but are not relevant to the subject of NER liveries.

Plate 9.15
Darlington North Road Works Plate from Class T2 No 2244.
(Private Collection)

Chapter 10

Locomotives for the Officers' Special Saloons

Aerolite

During Edward Fletcher's regime several small tank locomotives were assigned to hauling the saloons provided for the use of some of the principal officers of the company.

The locomotive chosen by Mr. Fletcher for his own use was the small 2-2-2T *Aerolite*, from the former Leeds Northern Railway (Fig 10.1); having been designed for working fast, light-weight passenger trains it was ideally suited for hauling the Locomotive Superintendent's special saloon. Unfortunately, *Aerolite* was badly damaged in a collision at Otterington in 1868 and was broken up soon afterwards.

The nameplates from *Aerolite* were transferred to a new 2-2-2T with inside cylinders and outside sandwich frames – 9in by 3in oak beams between ½in iron flitch plates – designed by Fletcher and built at Gateshead Works in September 1869 (Plate 10.1 and Fig 10.2); a pair of inside frames carried additional bearings for the crank axle. The new *Aerolite* had no number until October 1883 when McDonnell made it No 1428; it was renumbered 66 in February 1885.

No 66, the former *Aerolite*, was rebuilt in May 1886 with a larger boiler, new side tanks, cab and bunker of characteristic T W Worsdell design. (Fig 10.3) The existing oak beams, flitch plates and inner hornplates of the outside frames were retained, but the separate outer hornplates were replaced by single $^5/_8$in plates extending the full length of the locomotive between buffer beams; the leading wheels were moved 8in forward from their former position. The existing inside frames and most of the working parts underwent little alteration, however, at some unknown date the carrying wheels seen in Plate 10.1 had been exchanged for others with 10 instead of 12 spokes.

In November 1892 No 66 was rebuilt by Wilson Worsdell as a two-cylinder compound 4-2-2T, with enlarged side tanks and bunker (Plate 10.2 and Fig 10.4); the rear well tank was removed. The working pressure of the boiler was raised from 140psi to 160psi. The 1892 rebuilding was completed at the very time that Wilson Worsdell's new livery was about to supersede the previous ones. No 66 was painted in an intermediate style with neither a claret-coloured border around the tanks, cab and bunker, nor the later narrow black edge and white line at

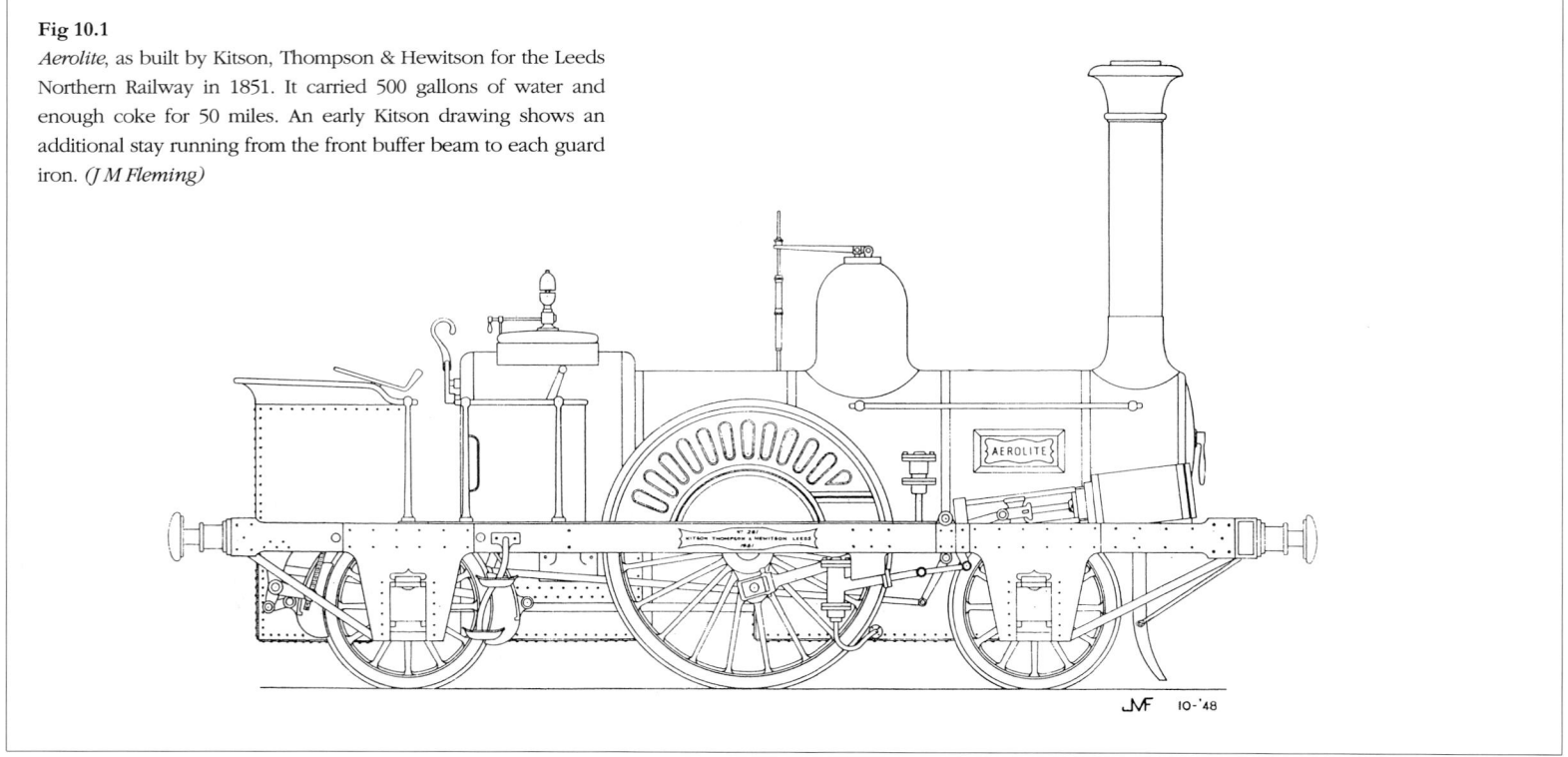

Fig 10.1
Aerolite, as built by Kitson, Thompson & Hewitson for the Leeds Northern Railway in 1851. It carried 500 gallons of water and enough coke for 50 miles. An early Kitson drawing shows an additional stay running from the front buffer beam to each guard iron. *(J M Fleming)*

Plate 10.1
Aerolite as built in 1869, retaining only the nameplates from its predecessor.

the top; the black bands of the panels were 2in wide with a $1/8$in white line on each side. The outside frames may perhaps have been painted claret colour in 1892 – available evidence is inconclusive – but later they were painted black with a $1/8$in vermilion line. By the late 1890s the side tanks bore the '1892' form of the NER armorial device between the letters 'N E' (without 'stops').

No 66 was again substantially altered in April 1902 when it was rebuilt as a 2-2-4T with new outside frames and an enlarged bunker. The large '1899' form of the NER armorial device was carried on the side tanks between the letters 'N E' (again without 'stops'); the panels on the tanks and bunker had a $1/8$in white line on the outer side of the black bands and a $1/4$in gold line on the inner.

The inside frames were renewed in September 1907 and the name *Aerolite* was restored to the much-rebuilt locomotive. New brass nameplates were affixed to the side tanks and the existing number-plates were replaced by new ones claiming (falsely) that the

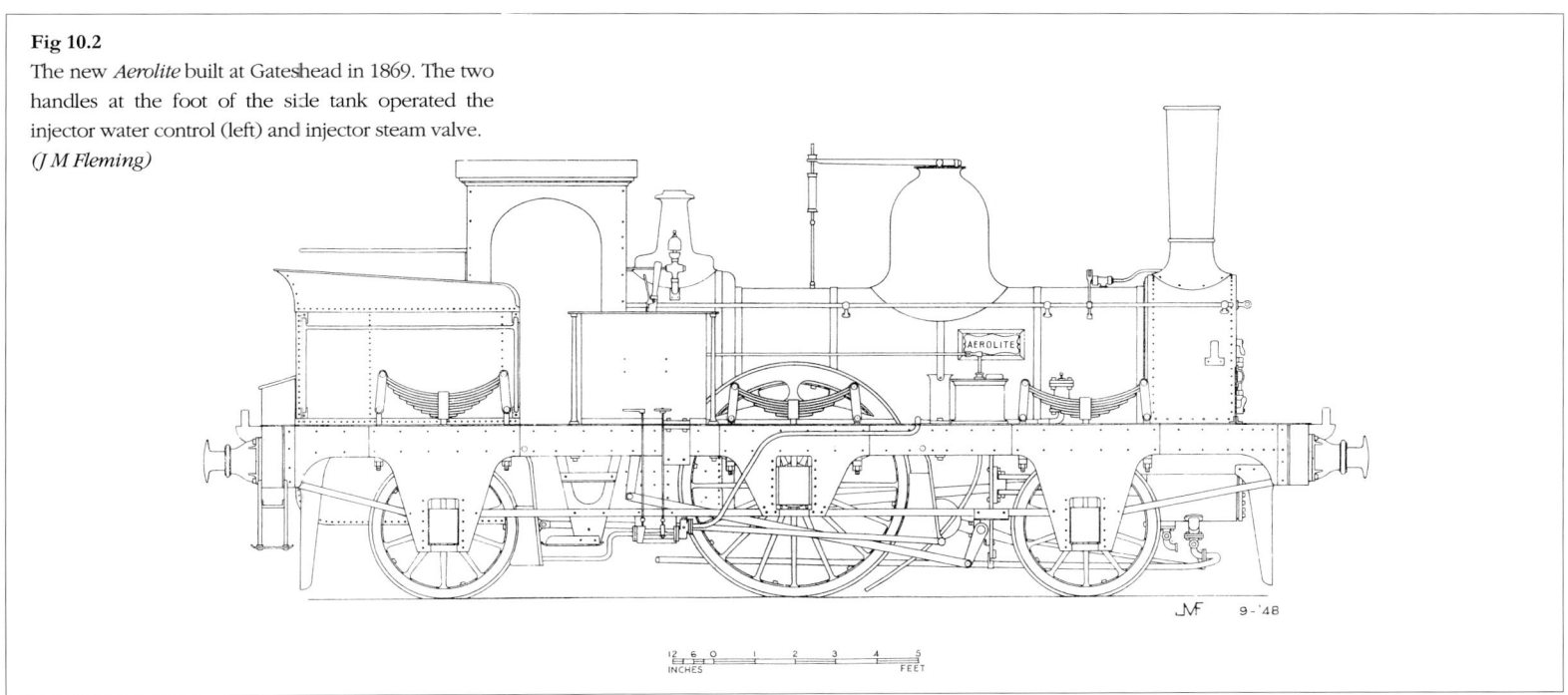

Fig 10.2
The new *Aerolite* built at Gateshead in 1869. The two handles at the foot of the side tank operated the injector water control (left) and injector steam valve.
(J M Fleming)

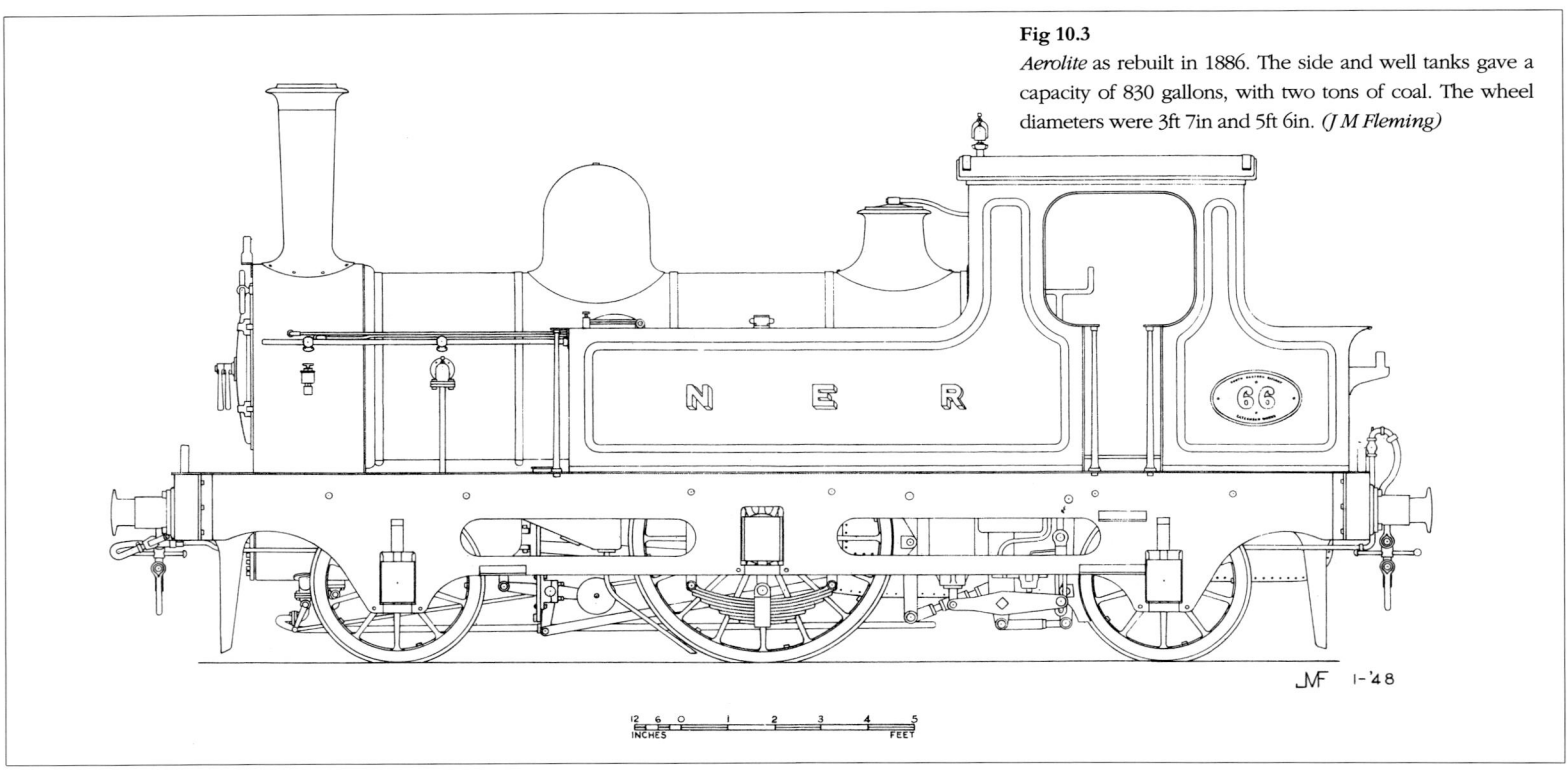

Fig 10.3
Aerolite as rebuilt in 1886. The side and well tanks gave a capacity of 830 gallons, with two tons of coal. The wheel diameters were 3ft 7in and 5ft 6in. *(J M Fleming)*

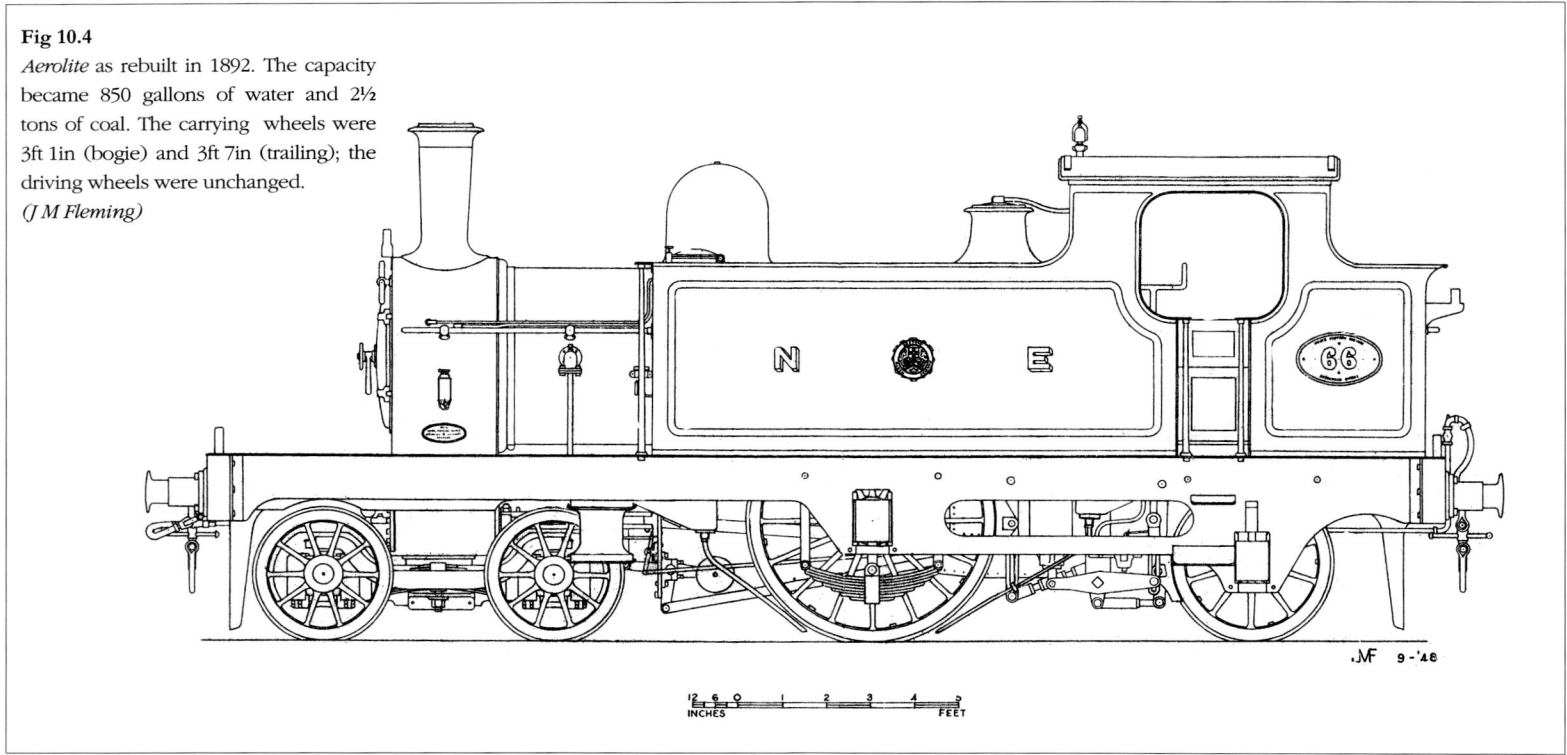

Fig 10.4
Aerolite as rebuilt in 1892. The capacity became 850 gallons of water and 2½ tons of coal. The carrying wheels were 3ft 1in (bogie) and 3ft 7in (trailing); the driving wheels were unchanged.
(J M Fleming)

locomotive had been built by Kitson & Co. in 1851; the background of the plates was vermilion. The nameplates were surmounted by the '1899' NER armorial device between 'NORTH' and 'EASTERN' in 4in gold letters arranged in the form of an arc of 12ft 8in radius. The black bands of the panels had a ³⁄₈in gold line on the inner side and a ¹⁄₈in white line on the outer. The outside frames were claret colour with a narrow black edge and ¹⁄₈in vermilion line. (Fig 10.5) The buffer casings and bufferbeam were vermilion, edged black and elaborately lined out in white and gold. The ornamental loops at the ends of the nameplates were copied from the plates on a 0-4-0 saddle tank named *Racey*, built by Black, Hawthorn & Co in 1895 for Sir Hedworth Wilkinson's Fulwell Limeworks, near Sunderland (Fig 10.6).

After its withdrawal, *Aerolite* was placed in the LNER Museum at York, the forerunner of the National Railway Museum.

Plate 10.2
Aerolite after its second rebuilding, by Wilson Worsdell in 1892. It is standing by the coaling stage at York shed, together with the Locomotive Superintendent's saloon, No 1661, built in 1871 and extended and put on bogies in 1904. Both locomotive and carriage have been preserved.

Fig 10.5
Aerolite as rebuilt in 1902, depicted following the renewal of the frames and restoration of the name in 1907. The water capacity had increased to 1,620 gallons, still with 2½ tons of coal. The bogie and trailing wheels of Fig 10.4 appear to have been re-used in the reversed ordering, but the driving wheel diameter had increased to 5ft 7¾in. *(J M Fleming)*

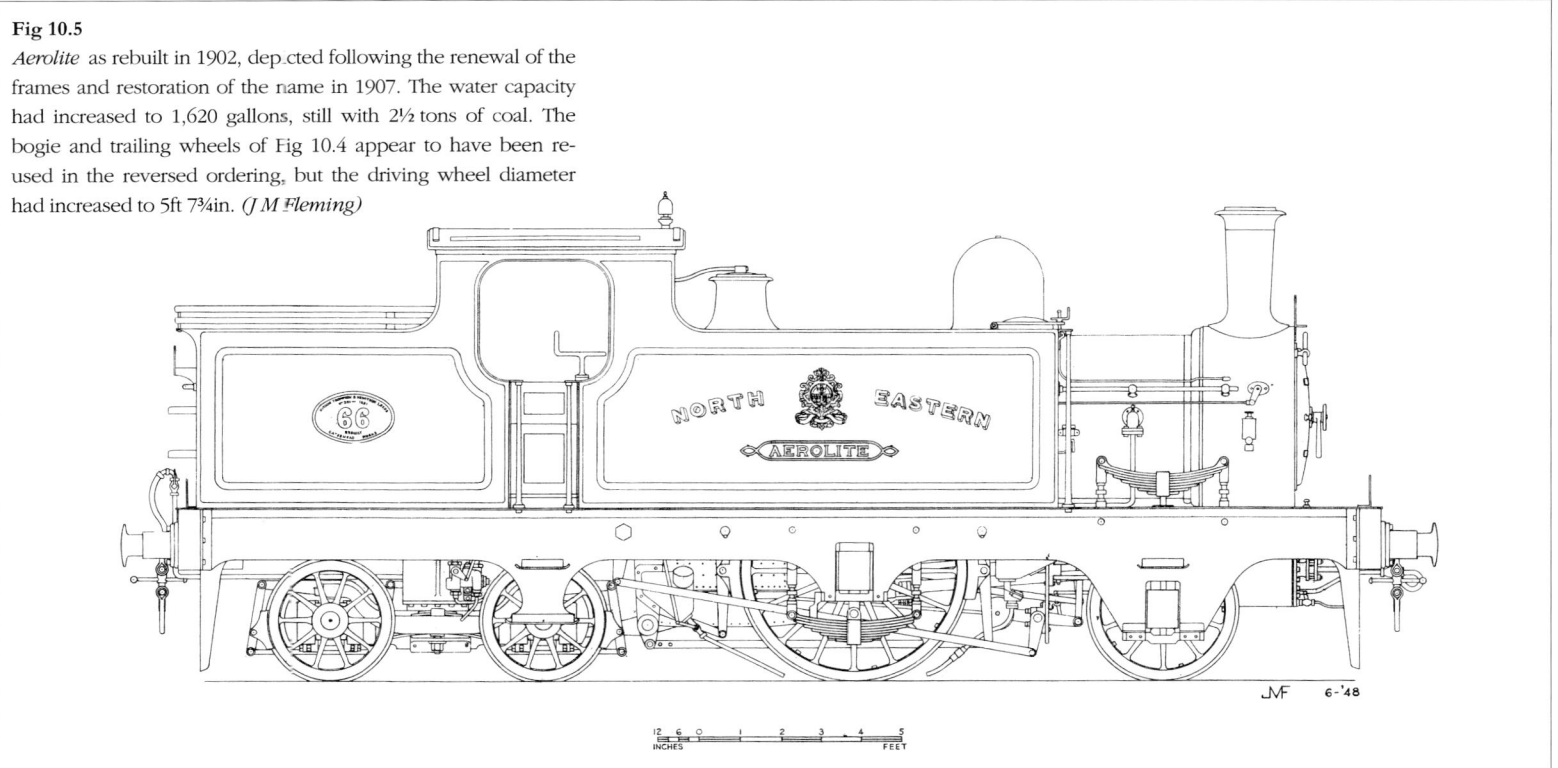

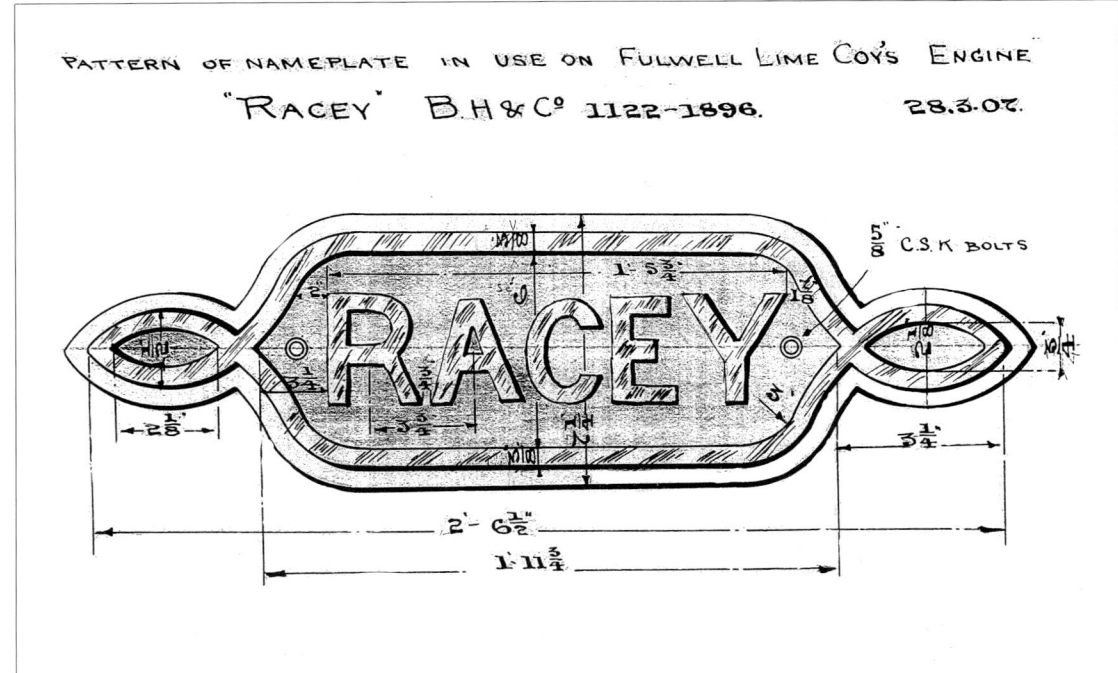

Fig 10.6a & b
Drawing by R H Inness of the nameplate used as a prototype for that installed on Aerolite in 1907, and an extract from the Gateshead Locomotive Drawing Office Register for 12 April 1907 where his drawings for *Aerolite*'s livery, nameplate and numberplate are listed as Nos 6663-5.

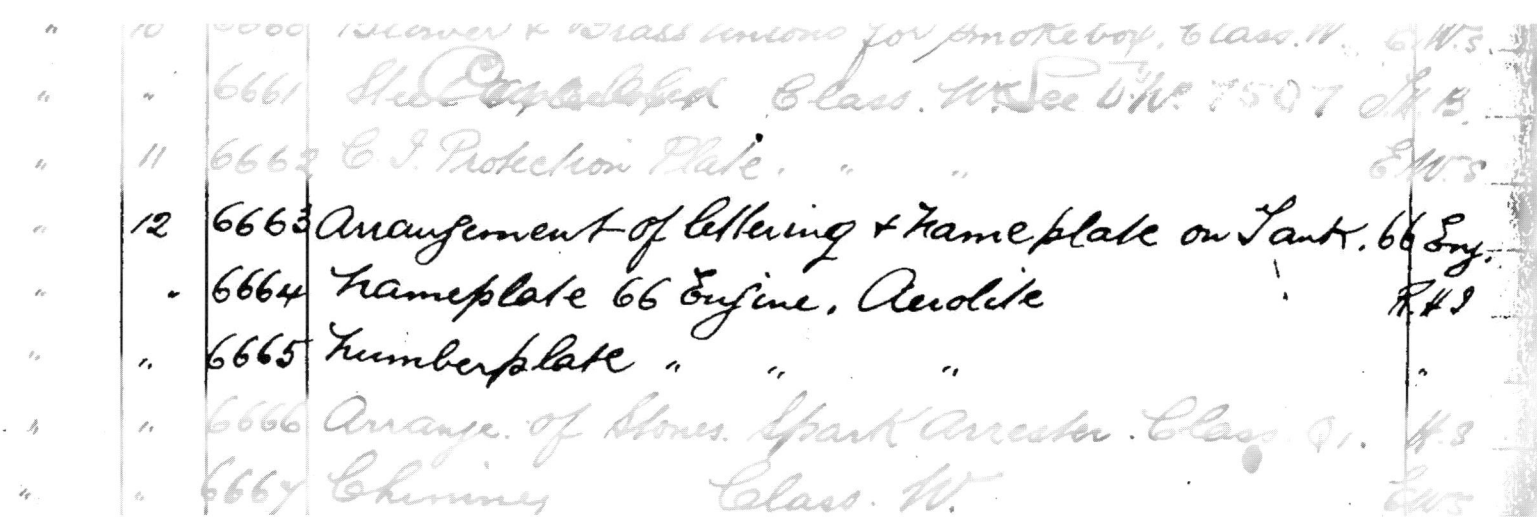

No 150

The *Aerolite* of 1869 had been preceded by a similar, but smaller, 2-2-2T No 150, built at Gateshead Works in February 1858. The wheel diameters were: leading 3ft 5½in; driving 5ft 6in; trailing 3ft 8in; the wheelbase was 6ft 0in + 8ft 7½in. Like *Aerolite*, No 150 had outside sandwich frames and a rear well tank below the bunker; it is known to have had a brass dome cover. E L Ahrons recorded that No 150 was stationed at Gateshead; it was renumbered 1718 in February 1885 and scrapped later in the same month. It was used by the Engineer's Department, Newcastle.

No 252

It is reputed that the diminutive 2-2-2T, No 252 (Plate 10.3), built at York Works in 1858, originated as a tender locomotive supplied to the York & North Midland Railway in 1840. No 252 had steam-operated brakes acting on the driving and trailing wheels and, although possessing only a weatherboard for the protection of the enginemen, it had doors enclosing the footplate – surely the earliest example of that usage.

The well tank held no more than 204 gallons of water, an additional supply being carried in a tank at one end of the inspection saloon at York, to which No 252 was invariably attached; a gong was fitted on the end of the saloon for signalling to the driver.

A larger bunker and an overall cab of York's peculiar design (Plate 10.4 and Fig 10.7) were fitted to No 252 in the 1870s. The cab was only 6ft 6in wide, and had glazed screens at each side which could be swung out to protect the driver when he leaned out to see ahead when propelling the saloon. A new 4-wheeled saloon was also built in the 1870s; it had a 568 gallon water tank below the floor, with a pipe connecting to the locomotive's well tank, and a speaking tube for communicating with the driver. This saloon, 28ft 10in over buffers

Plate 10.3
No 252, as built in 1858, with a carriage adapted for Officers' use.

Plate 10.4
No 252 equipped with a York Works cab, a later Fletcher chimney and a sandbox integrated with the driving-wheel splasher in a photograph taken before reboilering in 1885. A cord, communicating with a gong in the cab, can be seen hanging down from the saloon.

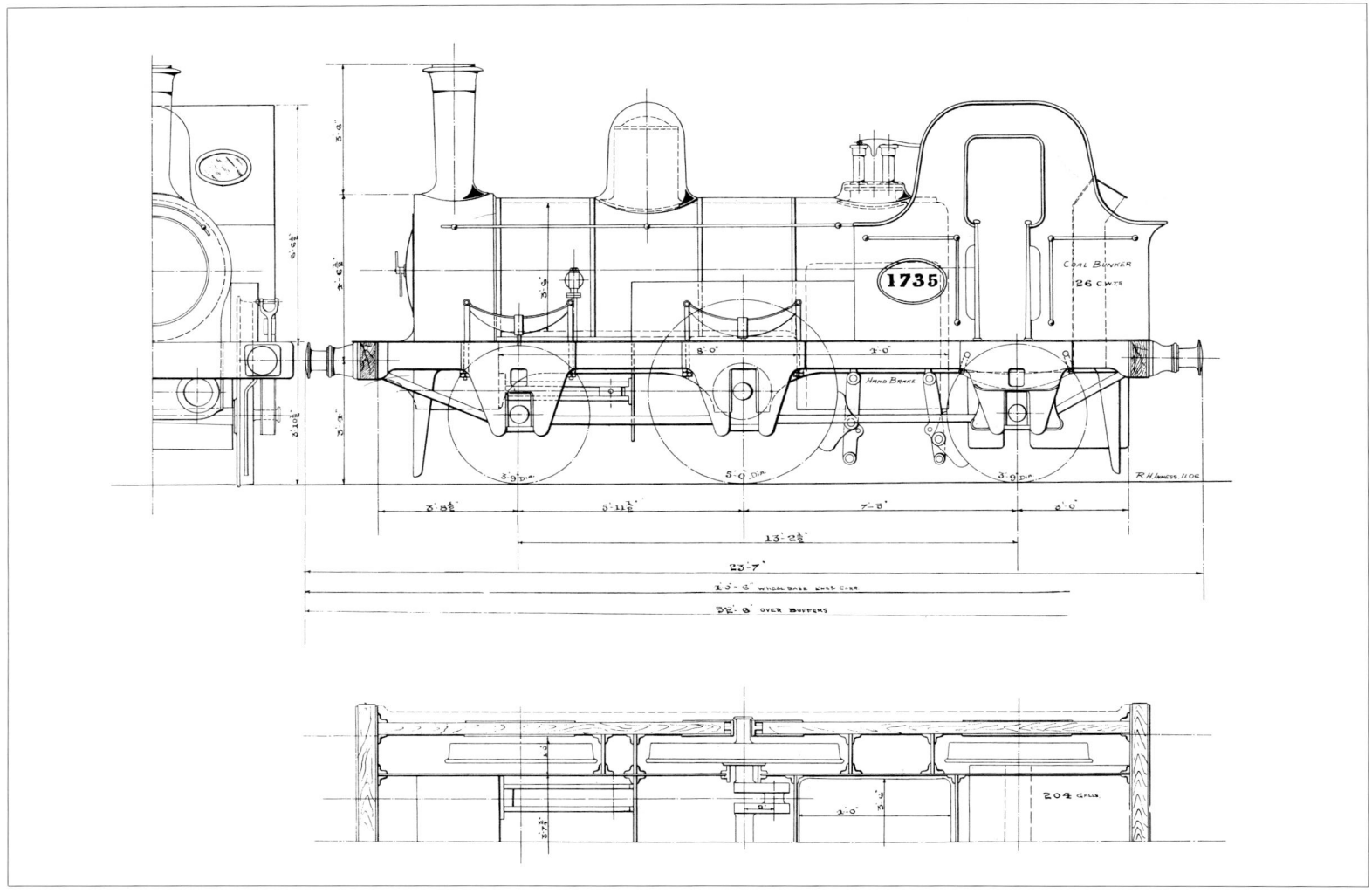

Fig 10.7
No 1735 (formerly No 252) equipped with a York Works cab. It was known as 'Mr. Copperthwaite's Engine': Harold Copperthwaite being the Southern Division Engineer, based at York, from 1888 to 1899. It is shown with its new boiler of 1885. The flat profile to the cab roof seems an unlikely change from the curve seen in Plate 10.4. *(R H Inness)*

and 16ft wheelbase, was perhaps No 853, built in July 1875. No 252 was reboilered in November 1885, renumbered 1931 in September 1891, 1735 in January 1894, and was withdrawn soon afterwards.

No 273

A fourth locomotive used for hauling an official saloon during Fletcher's regime was No 273, the Crampton-type 0-4-0T acquired by the York & North Midland Railway from the East & West Yorkshire Junction Railway in 1851. Built by E B Wilson & Co. in 1847 (Plate 10.5) for hauling fast, lightweight passenger trains, it was used latterly for the stores train and for the York Engineer's saloon. No 273 was shown at the Stockton & Darlington Railway Jubilee Exhibition in 1875.

Nos 190 and 1679

The 2-2-4 tank locomotives, Nos 190 and 1679, rebuilt at York Works in 1894, were conversions from 2-2-2 tender locomotives with a notable history. No 190 began as a very successful 2-2-2 express locomotive supplied by Robert Stephenson & Co to the York, Newcastle & Berwick Railway in January 1849 (Chapter 1, Fig 1.11). No 1679 originated as YN&BR No 77, one of Stephenson's pioneering 3-cylinder locomotives, built as a 4-2-0 in 1846 and rebuilt as a 2-2-2 in 1852-3 (Chapter 1, Figs 1.14 & 1.15). Both were rebuilt at York Works in 1881 with new outside frames and other parts (Fig 10.8 and Plate 10.7). In April 1891 No 77 was given a modified 'Class 1001' Worsdell boiler, with shortened barrel, and No 190 had been reboilered in the same way before their conversion to 2-2-4Ts in 1894. By then No 77 had been renumbered twice: to 1929 in September 1891 and then to 1679 in January 1894.

At first, both locomotives carried the initials 'N. E. R.' on the side tanks (Fig 10.9); later they bore the company's '1892' armorial device between 'N.' and 'E.' (Plate 10.6), the '1892' device being followed in due course by the '1899' version. Eventually the two locomotives bore the '1899' device between 'NORTH' and 'EASTERN'; the lining out was in gold and white, as on *Aerolite*, and they had brass-capped chimneys; the outside frames and the background of the numberplates were black.

Plate 10.5
No 273 in its original condition. It had a capacity of 225 gallons of water and one ton of coke.

No 694

For several years, Vincent Raven made use of a 2-4-0 tender locomotive of Class 686 – No 694 – to haul his saloon. It was one of those that had exchanged tenders with a McDonnell 4-4-0 and had been reboilered in 1902. No 694 was renumbered 11 in February 1908, the numbers 686-695 being required for new Class W 4-6-0Ts. No illustration of the locomotive as renumbered has been found, but it is known that its numberplate had a vermilion background, No 66 –

Fig 10.8
The former York, Newcastle & Berwick Railway No 190, as rebuilt with new frames, a Fletcher boiler and cab in 1881. *(E L Ahrons)*

Aerolite – being the only other locomotive to have that distinction after 1883. No 11 was not withdrawn until January 1911.

Class BTP

The Class BTP 0-4-4Ts were also chosen for hauling inspection saloons. In May 1902 the cab of No 955 was slightly enlarged and provided with side doors and a wooden roof: a brass-capped chimney was fitted and, alone among the BTPs, the '1892' heraldic device was

Plate 10.6

Class 190. No 1679 after rebuilding as a tank locomotive in 1894. It carries the 1892 armorial device. The locomotive was withdrawn in 1931. Hinged glass screens protected the driver and fireman while running tender first, since the rear windows could be obscured by coal.

Fig 10.9

No 190 rebuilt as a 2-2-4T at York in January 1894. It had a capacity of 990 gallons of water and two tons of coal. The wheel diameters were 4ft (leading), 6ft 6½in (driving) and 3ft 1¼in (bogie). *(J M Fleming)*

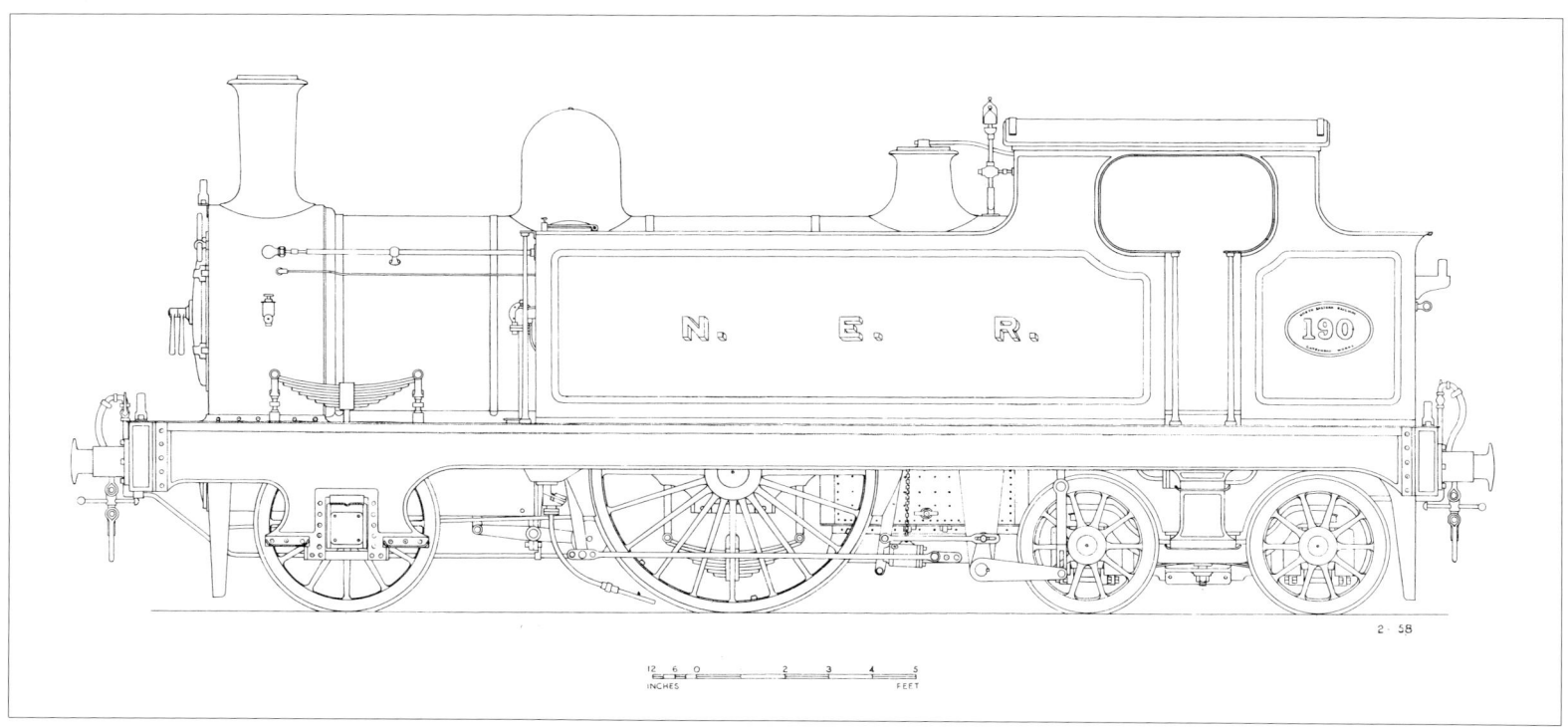

Plate 10.7
No 190 running with a Fletcher boiler and cab. *(R H Inness collection)*

displayed on the cab sidesheets; the original Fletcher number plate was retained, although repositioned. No 955 remained a 0-4-4T and was not confined to 'Officers' Special' workings.

No 957 was rebuilt as a 2-2-4T in April 1903, new leading and driving wheels replacing the 5ft 0in diameter coupled wheels; the existing bogie wheels were retained (Plate 10.8). Outside frames and side tanks were provided, the latter having a handrail along their full length and a beading along the top edge to match that on the bunker; the cab was altered like that of No 955. Three coal rails were added to the bunker, and a brass-capped chimney was provided. After a decade or more the side tanks were enlarged to bring them to the same height as the bunker, and the handrail was removed. The livery changes followed the same pattern as for Nos 190 and 1679, starting with the '1892' armorial device between 'N.' and 'E.'

Plate 10.8
No 957 as rebuilt in 1903.

Appendix 1

North Eastern Railway Locomotives in Preservation

Stockton & Darlington Railway

No 1 *Locomotion*	built 1825	exhibited at Darlington Railway Centre & Museum, North Road Station. (N)
No 25 *Derwent*	built 1845	exhibited at Darlington Railway Centre and Museum. (N)

North Eastern Railway

Class 66 *Aerolite*	built 1869	much rebuilt, exhibited at the National Railway Museum, York.
Class 1001 No 1275	built 1874	exhibited at Darlington Railway Centre & Museum. (N)
Class 901 No 910	built 1875	exhibited at Darlington Railway Centre & Museum. (N)
Class 1463 No 1463	built 1885	exhibited at Darlington Railway Centre and Museum. (N)
Class C No 876	built 1889	exhibited at Beamish Open Air Museum, County Durham.
Class H No 1310	built 1891	working on the Middleton Railway, Leeds.
Class M1 (later reclassified M) No 1621	built 1893	exhibited at the National Railway Museum, York.
Class Electric 1 No 1	built 1905	exhibited at the National Railway Museum, York.
Class T2 No 2238	built 1918	working on the North Yorkshire Moors Railway.
Class T3 No 901	built 1919	working on the North Yorkshire Moors Railway. (N)
Class P3 No 2392	built 1923	working on the North Yorkshire Moors Railway.
Class H No 985	built 1923	at Great Central (Nottingham) Ltd.
Class E1	built 1951	by British Railways as No 69023, working on the North Yorkshire Moors Railway.

(N) denotes a locomotive belonging to the National Collection but not stationed at the National Railway Museum. Working locomotives from the National Collection may be on loan to other preserved railways from time to time.

Appendix 2

The Duplicate List

In 1885, and again between January 1887 and January 1893, many old locomotives were replaced in capital stock by new, standard types and were transferred to a duplicate stock list, to be retained in service for a further period of months – or years! Initially, such locomotives retained their old numbers but as these were being re-used for new locomotives the numberplates were removed and the number painted on the cab side together with the letter A; an example of this is seen in Plate 1.2. Duplicate locomotives were then allotted numbers between 1701 and 1955, the new numbers being shown on vitreous enamelled iron plates of the same size, shape and colour as the standard brass numberplates. (Some early duplicate locomotives had their new number in gold transfers applied directly on the cab sides.)

The duplicate stock list was abolished on 1 January 1894, the surviving duplicate engines being allotted new numbers between 1656 and 1800 in the capital stock list. The new numbers were allocated on the basis of the age of the boilers, those engines with the newest boilers taking the lowest numbers. Engines thus restored to capital stock received new cast brass numberplates of standard pattern, with polished border and numerals but omitting the company's name and the building details.

In 1897 the duplicate list was revived and some old engines were renumbered 2051-5. A new duplicate series, starting at 2251, came into being in 1899; the cast brass number plates usually omitted the name of the company and never gave the date or place of origin. Later additions to the duplicate list in 1901, numbered above 2277, had enamelled iron numberplates.

Appendix 3

Wartime Loans

One of the First World War priorities was defence of the East Coast against naval bombardment. In 1915 two Class L 0-6-0Ts, Nos 544 and 545, were allocated to gun sites at Brotton and Hartley respectively, available for serving and manouevring the heavy rail-mounted guns. They were fitted with condensing gear, train heating connections and two Westinghouse pumps – one for the brakes and one for pumping feed water into the tanks from local sources, such as streams. During their military service they were painted grey, including their numberplates and other brasswork (Plate A3.1).

All 50 Class T1 0-8-0s were despatched in 1917 to France, where they worked under the control of the Railway Operating Division of the Royal Engineers. These locomotives were also painted grey, with the initials 'R O D' and the ROD number in large white characters on the tender, although the NER numberplates were retained. The ROD kept the NER numbers unless they duplicated those of locomotives from other sources, in which case they were increased by 5000. The locomotives were returned to the NER between February and July 1919, and in the following year their wartime services were commemorated by a brass replica of the Royal Engineers' insignia (a bursting grenade) and three chevrons attached to the cab side sheets (Plate A3.2 and Fig A3.1).

NER 0-6-0s were loaned to several home railways to cope with wartime traffic and arrears of locomotive maintenance afterwards. The Highland Railway was particularly hard pressed even during the

Plate A3.1
Class L No 544 equipped for war, with its two Westinghouse pumps.

first years of peace, and R H Inness recorded the despatch of eighteeen 0-6-0s of Class 398 to Perth on 19 November 1919, while, probably about the same time, he noted the loan of nine Class C to the Taff Vale Railway and three to the Maryport and Carlisle.

The NER itself was helped by the loan of 33 of the well-known ROD 2-8-0s, ordered by the War Department (WD) for service in France. These were a modified form of Robinson's superheated Class 8K, introduced on the Great Central Railway in 1911. Orders were placed by the WD from February 1917 and eventually 521 locomotives were built, though some arrived well after the need for their services had ended. From August 1919 they came under the newly-formed Ministry of Transport (MoT) and, having limited success in selling them, the Government offered the remainder on loan to the railway companies. The NER took 33, which retained their ROD numbers – which are given below. These were returned to the MoT in August 1921, Government control of the railways having at last ended on the 15th of that month. Ten of these locomotives were among the 273 RODs later purchased by the LNER.

ROD Nos 1734-37, built by Robert Stephenson & Co, supplied to NER on 7 November 1919 new from strorage at Barnbow, Leeds.

ROD Nos 2073-76, built by North British Locomotive Co, supplied to NER on 15 October 1919 new from Barnbow.

ROD Nos 1729-32, supplied directly by Nasmyth Wilson & Co, Patricroft, in January-February 1920.

ROD Nos 1747-9, supplied directly by Robert Stephenson & Co in February (No 1747) and May 1920.

ROD locomotives supplied from storage at Tattenham Corner, in Surrey, after service in France, were Nos 1940, 44, 50, 55, 58, 60, 61, 64, 66, 67, 77, 82, 83, 87, 89, 90, 96, 97.

Most of the NER RODs returned in 1921 went into storage at former wartime Ministry of Food sidings at Royds Green on the East & West Yorkshire Union Railways.

Footnote: Details of RODs supplied to the NER are taken from J W P Rowledge, *Heavy Good Engines of the ROD, Volume 1*, published by Springmead Railway Books, 1977.

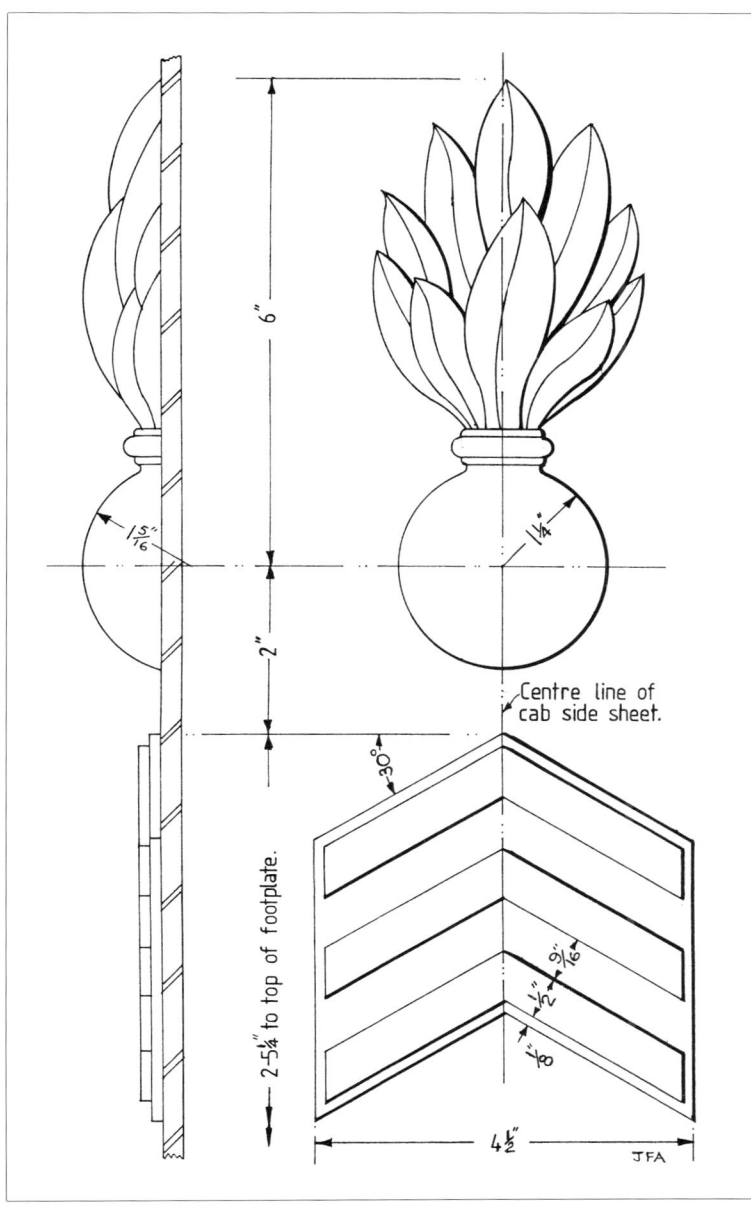

Plate A3.2
Class T1 No 1704, bearing its brass grenade and chevrons above the Gateshead Works plate. *(J F Mallon collection)*

Fig A3.1
The grenade and chevrons applied to class T1. *(J F Addyman)*

Appendix 4

Some Dimensions

Transverse Dimensions

To assist with modelling, in the absence of many front elevation drawings, two of the principal transverse dimensions are given for the later classes. These are the width over footplate and width over cab.

Class		Width		Class		Width		Class		Width	
LNER	NER	Footplate	Cab	LNER	NER	Footplate	Cab	LNER	NER	Footplate	Cab
	BTP	8ft 0in	7ft 10in	Y8	K	7ft 8in	7ft 1in	Q5	T & T1	8ft 6in	7ft 5in
J74		8ft 0in	7ft 10in	J73	L	8ft 4in	7ft 9in	Q6	T2	8ft 6in	7ft 5in
A2	Pacific	8ft 6in	7ft 10in	D17/1	M	8ft 0in	7ft 2in	Q7	T3	8ft 6in	7ft 5in
F8	A	8ft 4in	7ft 9in		3CC	8ft 0in	7ft 2in	N10	U	8ft 4in	7ft 9in
N8	B	8ft 4in	7ft 9in	N9	N	8ft 4in	7ft 9in	C6	V	8ft 6in	7ft 5in
J21	C	8ft 0in	7ft 2in	G5	O	8ft 4in	7ft 9in	A6	W	8ft 6in	8ft 1in
	D 2-4-0	8ft 0in	7ft 2in	J24	P	8ft 0in	7ft 2in	T1	X	8ft 6in	8ft 4in
A1, A8	D 4-4-4	8ft 6in	8ft 4in	J25	P1	8ft 0in	7ft 2in	A7	Y	8ft 6in	8ft 4in
J71	E	8ft 0in	7ft 7in	J26	P2	8ft 0in	7ft 2in	C7	Z	8ft 6in	7ft 5in
J72	E1	8ft 0in	7ft 7in	J27	P3	8ft 0in	7ft 2in	C8	4CC	8ft 9in	8ft 0in
D22	F	8ft 0in	7ft 2in	D17/2	Q	8ft 0in	7ft 2in	X1	66	8ft 0in	7ft 6in
D23	G	8ft 0in	7ft 2in	D20	R	8ft 0in	7ft 3in	X2	957	8ft 0in	not known
Y7	H	7ft 8in	7ft 1in	D21	R1	8ft 6in	7ft 5in	X3	190	8ft 4in	7ft 9in
J78	H1	7ft 8in	7ft 1in	B13	S	8ft 6in	7ft $3^3/_8$in	X3	933	7ft 4in	7ft 1in
J79	H2	7ft 8in	7ft 1in	B14	S1	8ft 6in	7ft 5in	J77	290	8ft 0in	7ft 6in
	I	8ft 0in	7ft 2in	B15	S2	8ft 6in	7ft $3^3/_8$in				
	J	8ft 0in	7ft 5in	B16	S3	8ft 6in	7ft 5in				

Plate A4.1
Rear view of a Fletcher tender showing the narrow tank squeezed between the springs. Note the side chains flanking the coupling hook. It has acquired coal rails and is seen in Wilson Worsdell livery, though the lines of rivets indicate why in Fletcher livery the tender was normally lined out in three panels. The locomotive is Class 1440, No 1448, built in 1878 but shown with the Worsdell boiler fitted in 1891. The location is Gateshead.
(J F Mallon collection)

Index

NER Locomotive Classes:

8	90	1162	79	P	113, 116-7, 125
11	72	1196	51	P1	116-7
13	25-6, 47	1238	67, 79	P2	126-7
25	28-9, 45-6	1265	67, 79	P3	126-7, 137-9
30	64-5	1350	64-5, 110	Q	114-16
36	81-2	1395 (Class 398)	54	Q1	115-16
38	86-7, 90	1440	58-9, 72	R	119-21, 128
40	79-80	1463	79, 89-90	R1	127-8
41	26-8, 63	BTP	58-60, 80-1, 110, 117, 139, 152	S	119-22, 126, 136-7
44	84	A	91-93, 98, 108, 113	S1	121-2, 126, 137
59	85-87	B	98, 100, 111	S2	122, 134-7
93	26-7	B1	92, 98	S3	139-40, 143
120	63	C	92-100, 107-8, 116, 157	S/07	122, 134
124	83-4, 110	C1	92, 96, 113-16	T	124-25, 134
162	27	D (2-4-0)	92-99, 114	T1	124-7, 136, 156-7
287	83-4	D (4-4-4T)	135, 137	T2	134, 143
290	80, 116-8, 139	E	95-98, 108, 111, 117, 119	T3	108, 130, 139-40
398	54, 61-3, 68-9, 81-3, 157	E1	111, 116-9, 135-6	U	116-8
447	27-9	F	97-100, 103	V	124-8
476	64-6	F1	98-100, 114, 126	V/09	122, 127-8
492	44-5	G1	98-9, 126	W	128-30
544	44-6	H	104-108, 115, 124	X	129-30
577	48-9	H1	105-106, 115	Y	132-3
603	83	H2	115-117	Z	122, 128, 132-9, 143
675	55-6	I	99-103, 126	Z1	132, 136-7
686	54-5	J	99-103, 108, 114, 126	4-6-2	141-2
708	58-61	K	106-7	3CC	114-5
901	55-8, 72, 88, 128	L	108, 156	4CC	122-4, 138
964	63-4, 110-3	M	100, 108-10, 114	Electric 1	130
964A	111-3	M1	100, 108-10, 114-5	Electric Freight	137-8
1001	53, 67-8	N	110-11	Electric Express	141
1068	67	O	112-4, 117, 130	*Aerolite*	144-8

NER Locomotive Liveries:

W Bouch	66
E Fletcher	12, 13, 46, 54, 57, 61-4, 72, 85, 158
A McDonnell	87-9
V Raven	136-42
H Tennant	90
T W Worsdell	84, 88, 92-3, 96-7, 100, 107-11
W Worsdell	59, 102-3, 108-10, 114-5, 121-27, 131, 144-46, 158

NER Constituent Railway Companies:

Blyth & Tyne	69-71
Brandling Junction	3
Cawood, Wistow & Selby	116-7
East & West Yorkshire Junction	14, 150
Great North of England	1
Great North of England, Clarence & Hartlepool Junction	4
Hartlepool Dock & Railway	4
Hull & Selby	13-15
Leeds & Selby	5, 13
Leeds & Thirsk	17-18
Leeds Northern	1, 18-20, 22, 144
Londonderry	131
Malton & Driffield Junction	1
Newcastle & Berwick	5
Newcastle & Carlisle	29-31
Newcastle & Darlington Junction	1, 19, 22
Newcastle & South Shields	5, 46
Pontop & South Shields	5
Stanhope & Tyne	5
Stockton & Darlington	18, 31-42, 50
West Hartlepool Harbour & Railway	42-3
Whitby & Pickering	14
York & Newcastle	1, 4
York & North Midland	1, 13-18, 22, 148, 150
York, Newcastle & Berwick	1, 4-13, 22, 46, 49, 150-1

NER Workshops:

Darlington	20, 50-3, 66-9, 72, 79-83, 86-90, 98, 126, 130, 143
Gateshead	19-20, 23-9, 49-50, 54-8, 61-6, 72, 80-1, 85, 89, 98, 104, 128, 130, 144, 148
Hartlepool	42-3
Leeds	20, 58
Percy Main	69-71
Shildon	31-7, 50-1
York	20, 23, 58, 85, 117, 148-50

Locomotive Builders:

Avonside Engine Co	68
Beyer Peacock & Co	54-5, 58
Black, Hawthorn & Co	63, 146
Coulthard, J	8-9
Dubs & Co	61, 68
Fenton, Murray & Jackson	14
Fossick & Hackworth	34, 43
Gilkes, Wilson & Co	37-40, 68
Hackworth, Timothy	6, 7, 33
Hawthorn, R W & Co	2-11, 19, 26-32, 38-9, 44-7, 52-3, 58, 61, 64, 80, 86, 88
Hopkins, Gilkes & Co	52
Hudswell Clarke & Co	47-8
Kirtley & Co	13
Kitching, W A & Co	34-7
Kitson & Co	20, 23
Kitson, Thompson & Hewitson	17-19, 144
Longridge, R B & Co	2, 3, 15, 18
Manning Wardle & Co	24-5, 48-9
Nasmyth, Wilson & Co	157
Neilson & Co	55-61, 80
North British Locomotive Co	157
Sharp, Stewart & Co	54, 61-2
Shepherd & Todd	13-14
Stephenson, Robert & Co	1-11, 14-22, 26, 28, 31, 38-40, 44, 47, 50-1, 55, 58, 60-3, 79, 86, 89, 150, 157
Thompson Bros	31
Todd, Charles & Co	18
Wilson, E B & Co	8, 12, 14-6, 18, 20-4

Personalities:

von-Borries, August	91, 96
Bouch, William	31, 33-40, 44, 50-2, 66, 68, 79
Brown, John	iv, 88
Brown, William G	9, 13, 46
Cabry, Thomas	15
Copperthwaite, Harold	150
Cudworth, James I'Anson	1
Fletcher, Edward	iii, 20, 23-9, 44-6, 54, 58, 66, 69, 72, 82, 132, 144
Gilkes, Edward	31
Gilkes, Oswald	31
Graham, George	89
Graham, John	37
Gray, John	14-5
Hackworth, Timothy	31
Harrison, Thomas Elliot	29
Heppell, George	124
Inness, Richard H	iii, 2, 34, 68, 148, 157
Joy, David	15
Kendal, Ramsay,	72
Kendall, William	71
MacLean, John S	iii, 15, 89
McDonnell, Alexander	85, 86, 88, 144
Raven, Vincent Litchfield	119, 132, 151
Smith, Walter Mackersie	98, 114, 122, 124
Tennant, Henry	89-90
Thompson, W B	54, 61
Watson, G M	12, 31
Worsdell, Thomas William	iii, 15, 91-2, 95-98, 107, 144
Worsdell, Wilson	89, 106, 108, 126, 132, 144, 147

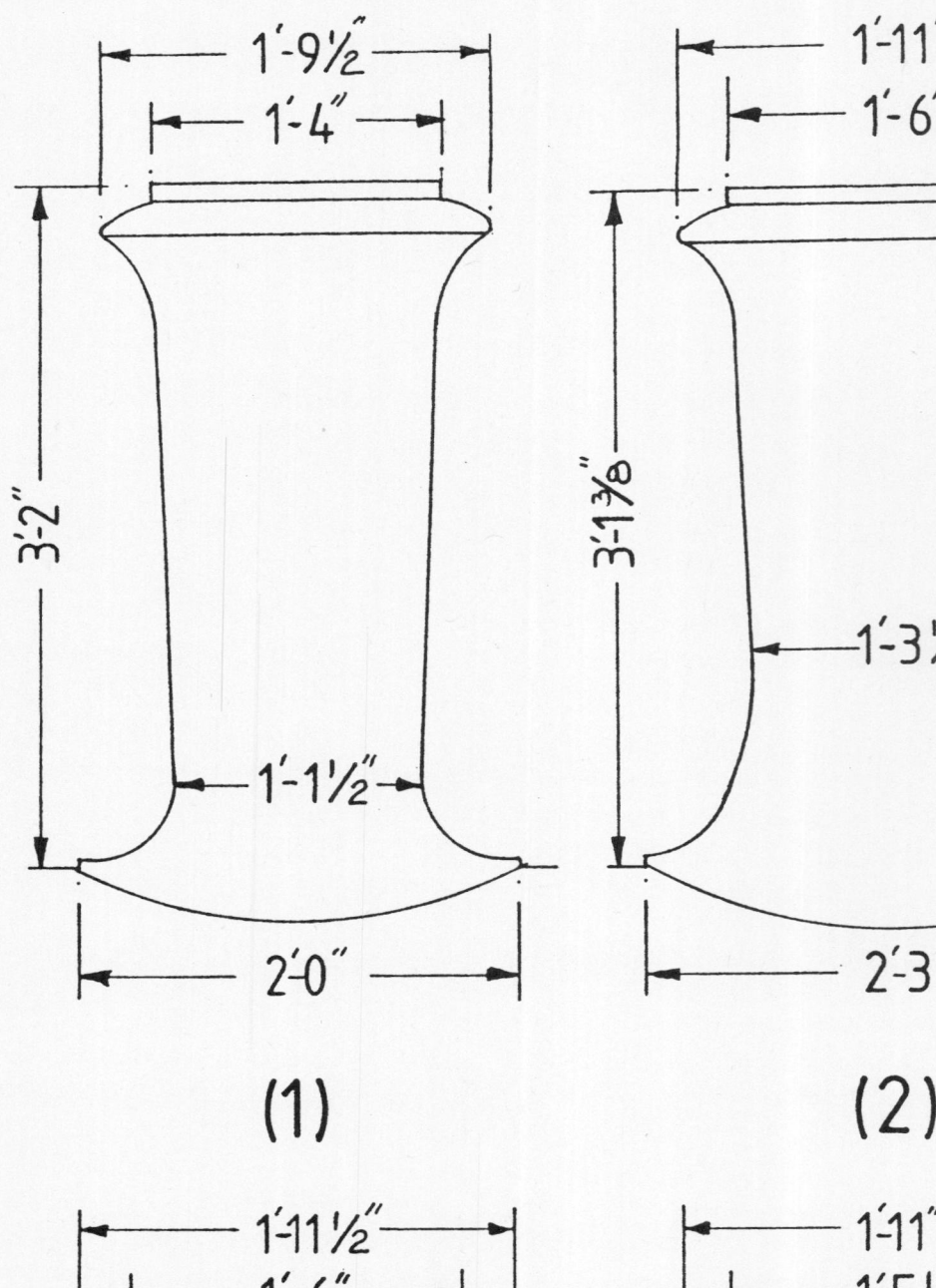
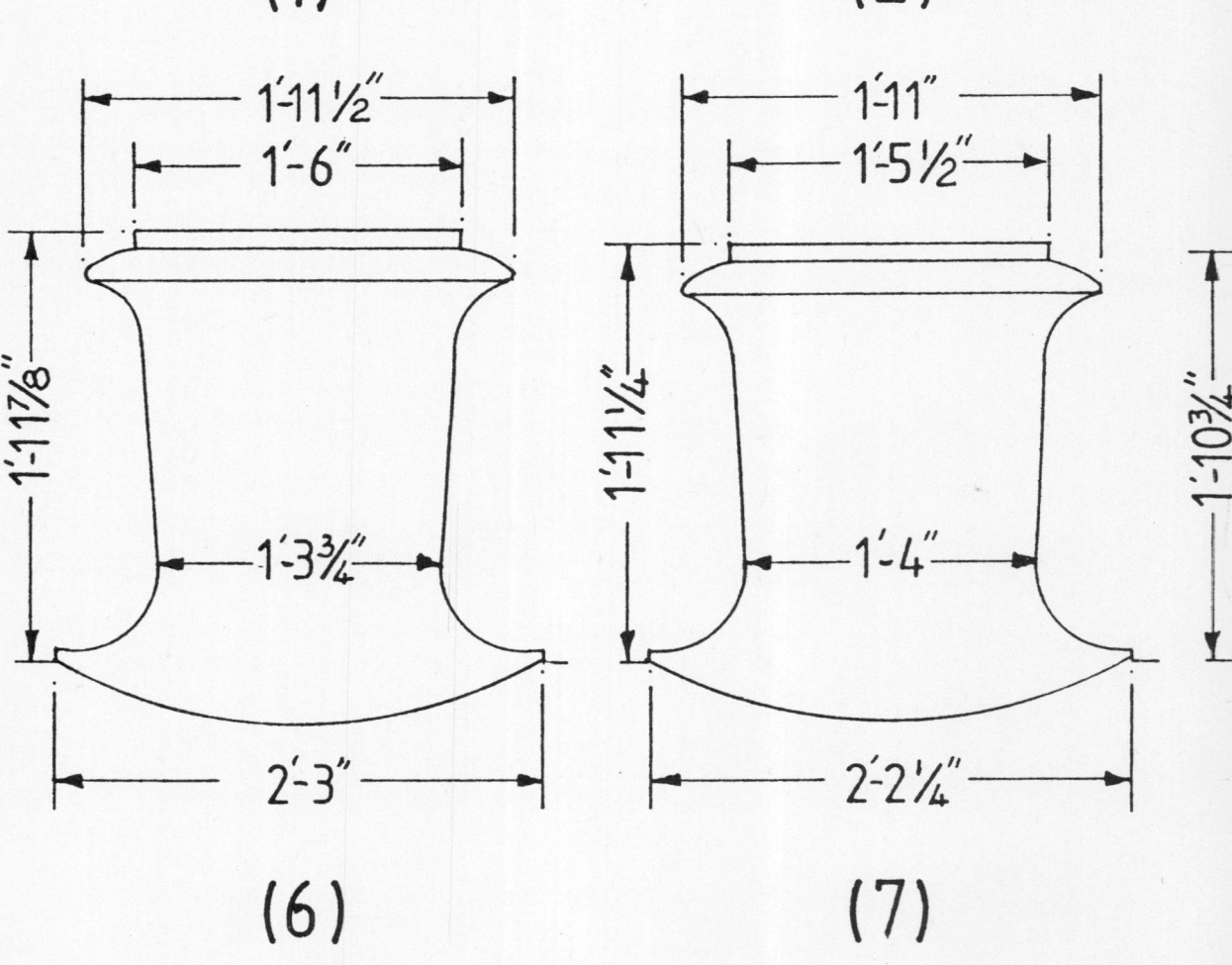

North Eastern Railway Chimneys

Key of classes to which they were fitted:

1 E, E1, G, H (reduced in height by 2 inches)
2 A
3 B, C, L, N, O, P, P1, U, 290
4 K
5 M, Q
6 S1 (shortened by 3 inches), X
7 S, T, T1, T2
8 R
9 Y
10 P2, P3